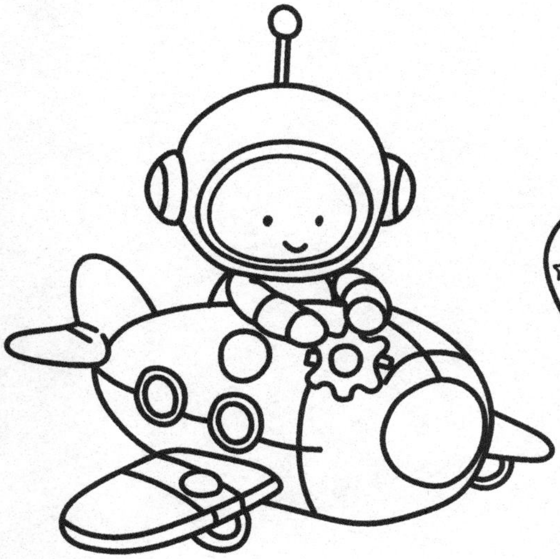

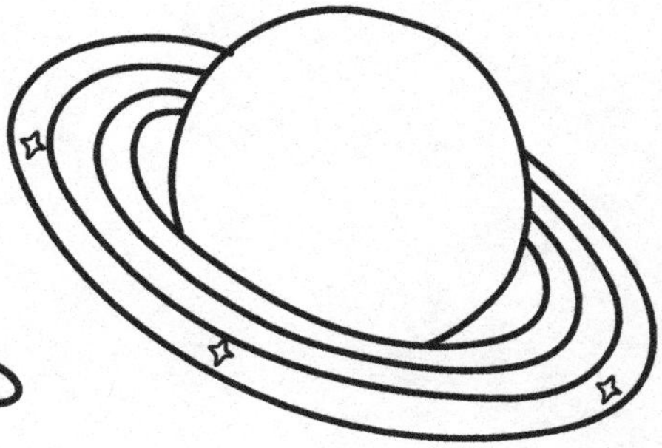

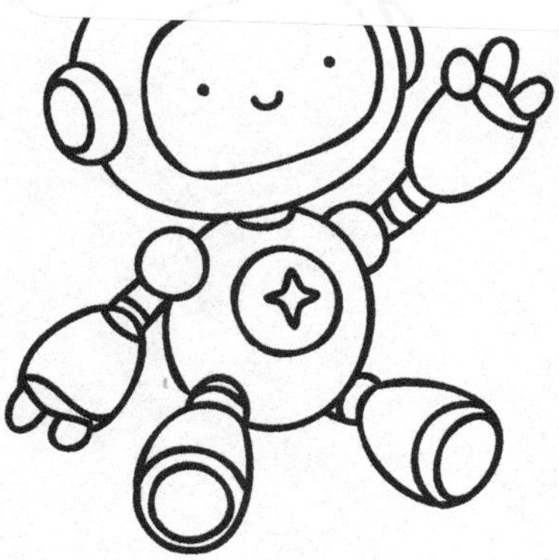

COLOR WITH NIA & EMERY

ROBOTS, PLANETS & GALAXIES

Ina Farrell

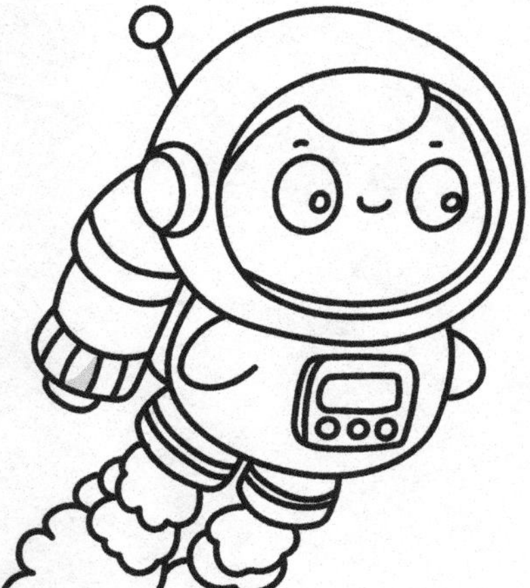

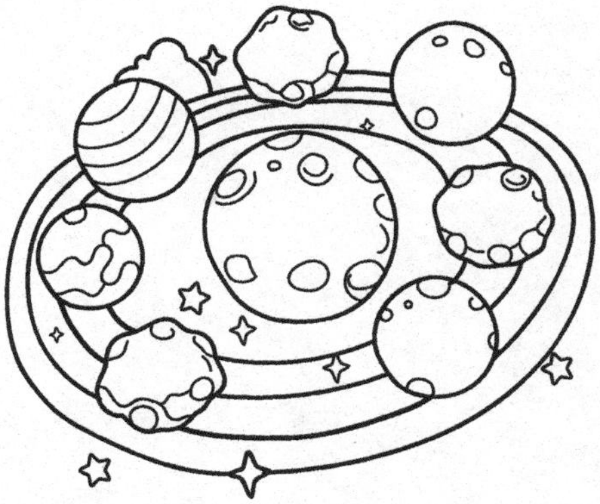

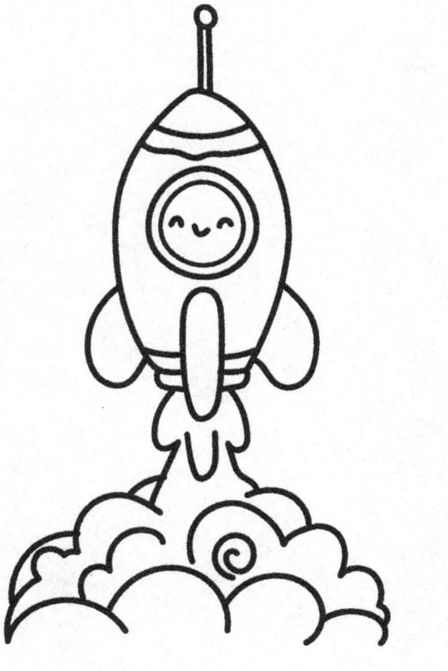

Info@workwithina.com

A SPECIAL MESSAGE FROM NIA & EMERY

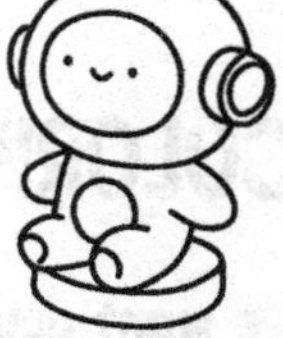

Hi there, explorer!

Welcome to "ROBOTS, PLANETS & GALAXIES!" This coloring book is all about the wonders of space, the beauty of the heavens, and the power of science and technology. From glowing planets and shining stars to robots and incredible spacecraft, there's so much to discover and color!

Have you ever looked up at the night sky and wondered about the worlds beyond our own? Space is a **masterpiece of creation, filled with mysteries waiting to be explored.** Scientists, engineers, and astronauts work every day to help us learn more about the universe—and now, you get to bring it all to life with your colors!

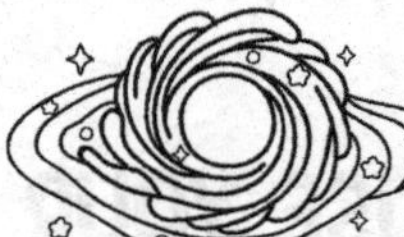

"WHEN CURIOSITY MEETS CREATIVITY, NEW WORLDS OPEN UP."

We've designed every page with bold lines and simple designs so that coloring is fun and easy for all ages. So grab your crayons, markers, or colored pencils and let's explore the incredible world of space and technology together!

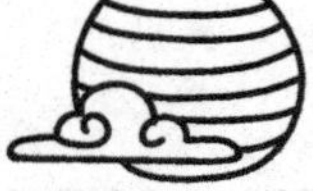

SHARE YOUR STELLAR CREATIONS & INSPIRE OTHERS!

We'd love to see your colorful creations! ⊠ Share your finished pages on Amazon and social media to inspire others to join in the fun. Your creativity helps us continue making books that bring **joy, imagination, and relaxation** to kids and grown-ups everywhere.

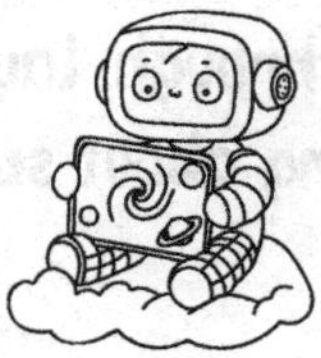

Every review brings us one step closer to achieving our dream of going to college!

Don't forget to tell your friends, family, and everyone else about the fun you're having with "ROBOTS, PLANETS & GALAXIES!" We can't wait to see your amazing coloring skills in action.

Thank you for exploring the universe with us!

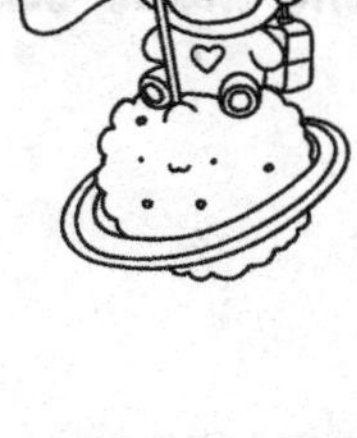

With love,
Nia & Emery

COLORING TIPS FOR ROBOTS, PLANETS & GALAXIES!

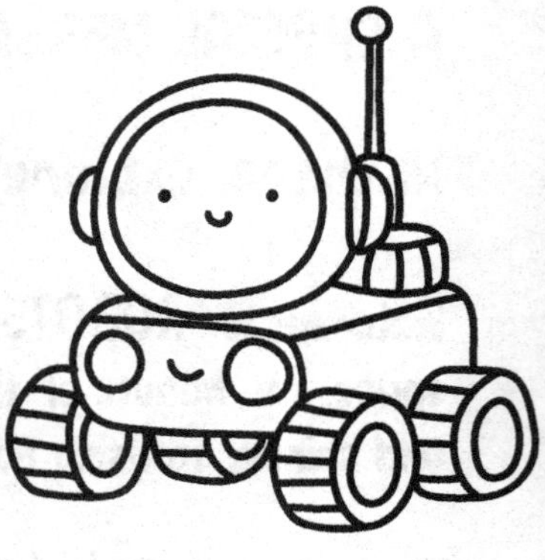

We want you to have the most relaxing and enjoyable time coloring, so here are some helpful tips to get you started:

1. ONE-SIDED FOR A REASON!

Each page is printed on one side to help prevent any bleed-through when using markers or other coloring tools. Just slide a thick sheet of paper under the page you're coloring to keep the other pages safe! (See illustration on the next page)

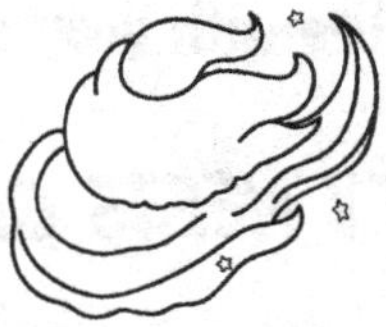

2. USE YOUR FAVORITE COLORING TOOLS!

Whether you're using crayons, colored pencils, markers, or gel pens, have fun bringing the wonders of space and technology to life! Test your tool on a small section or on the Test Color Page we included to make sure it's perfect for your masterpiece.

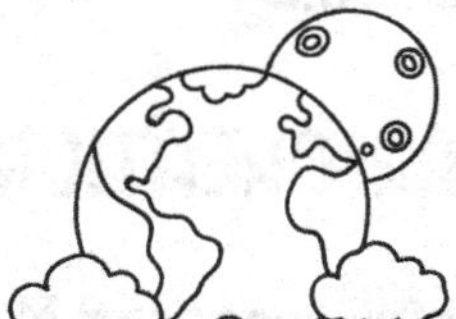

3. PERFECTLY SIZED FOR FUN!

At 8.5 x 8.5 inches, this book is designed for easy handling and coloring. It's the perfect size to flip through, lay flat, and color anywhere you like—whether you're at home, on a road trip, or dreaming about space exploration.

4. LET YOUR CREATIVITY SHINE!

From glowing planets to fun-to-color robots, each page is a chance to create something brilliant and inspiring. Take your time, relax, and make each picture your own with colors that spark your imagination.

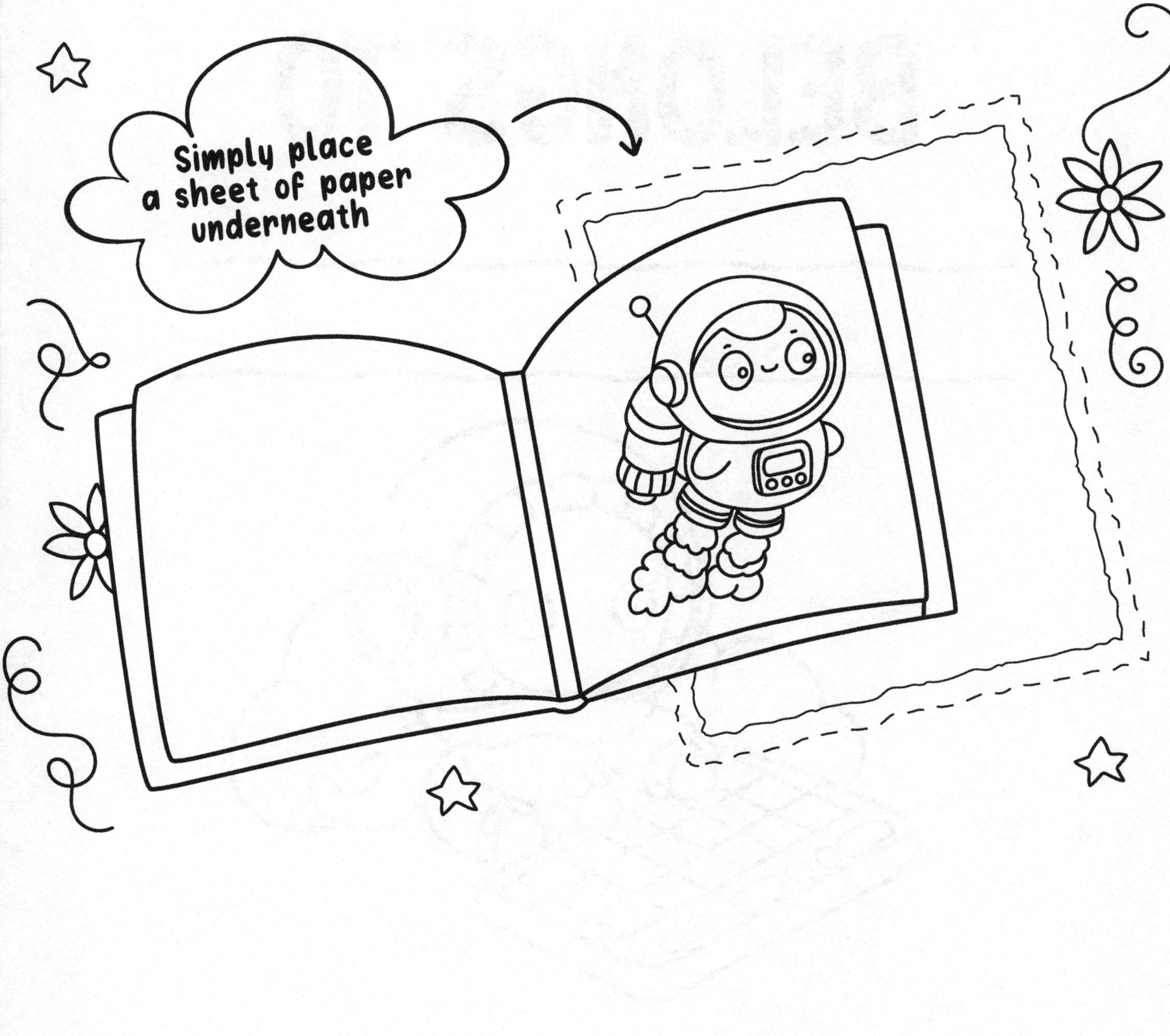
Simply place a sheet of paper underneath

THIS BOOK BELONGS TO

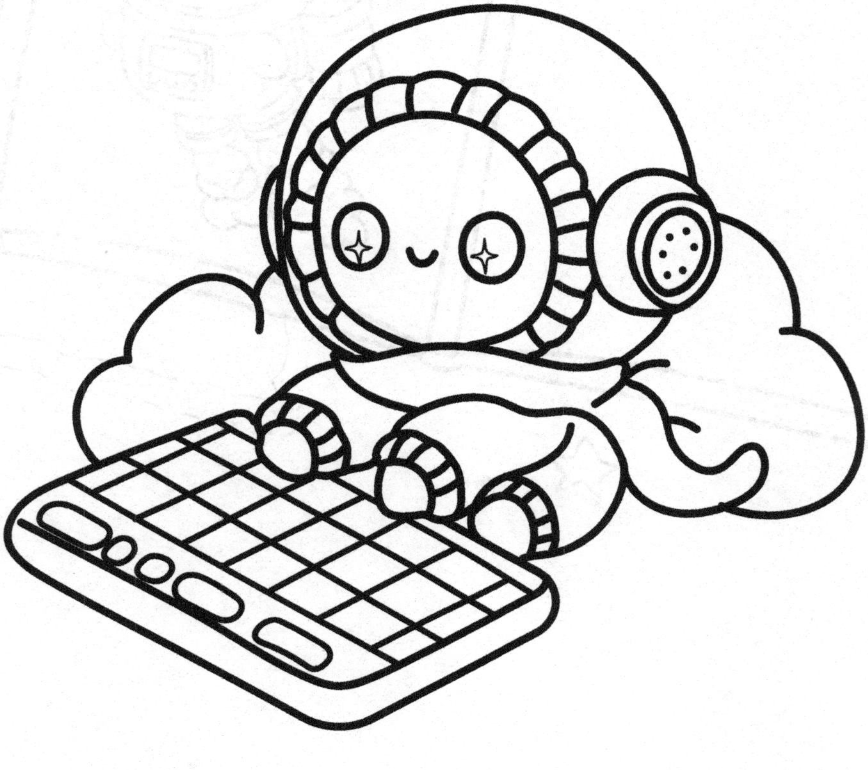

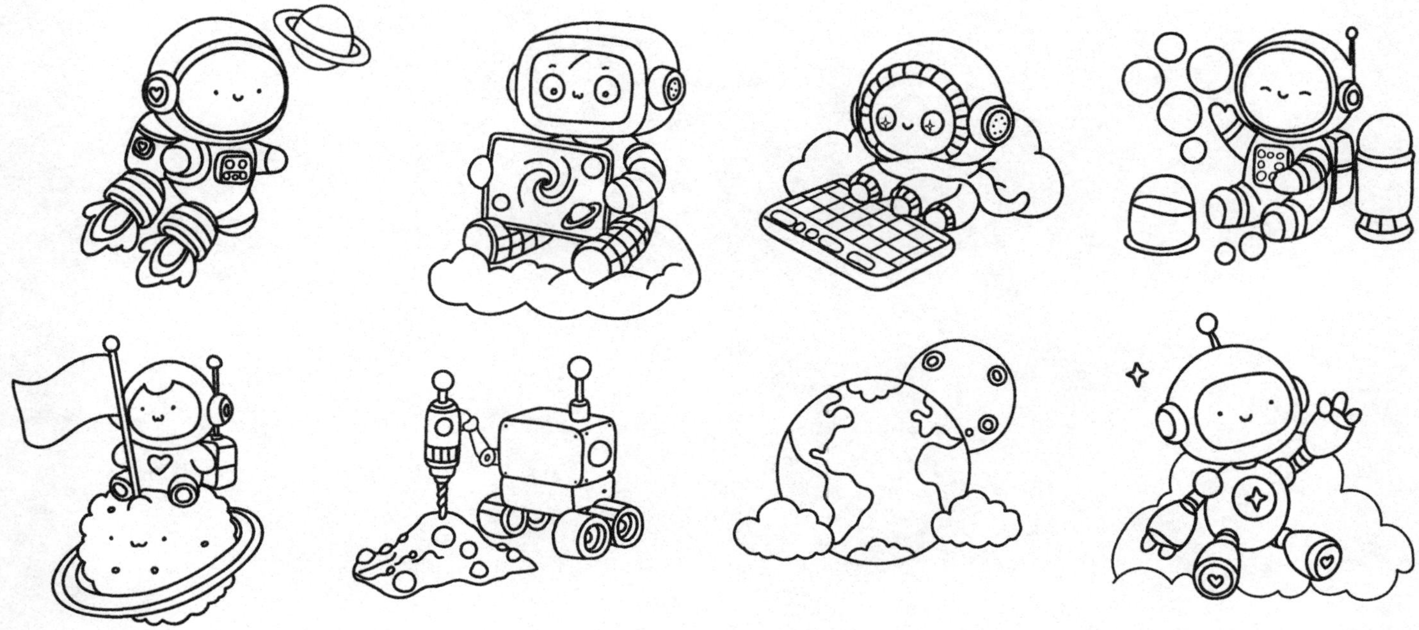

TEST COLOR PAGE

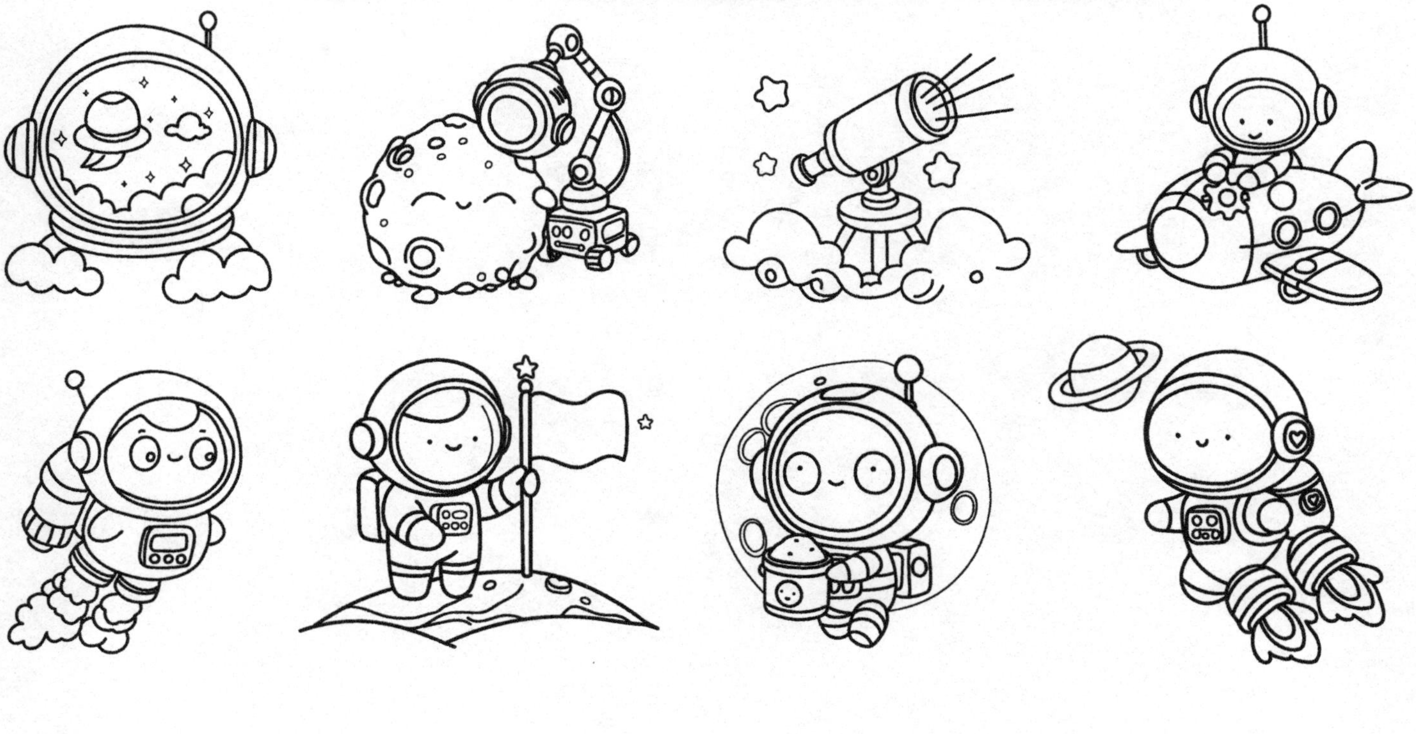

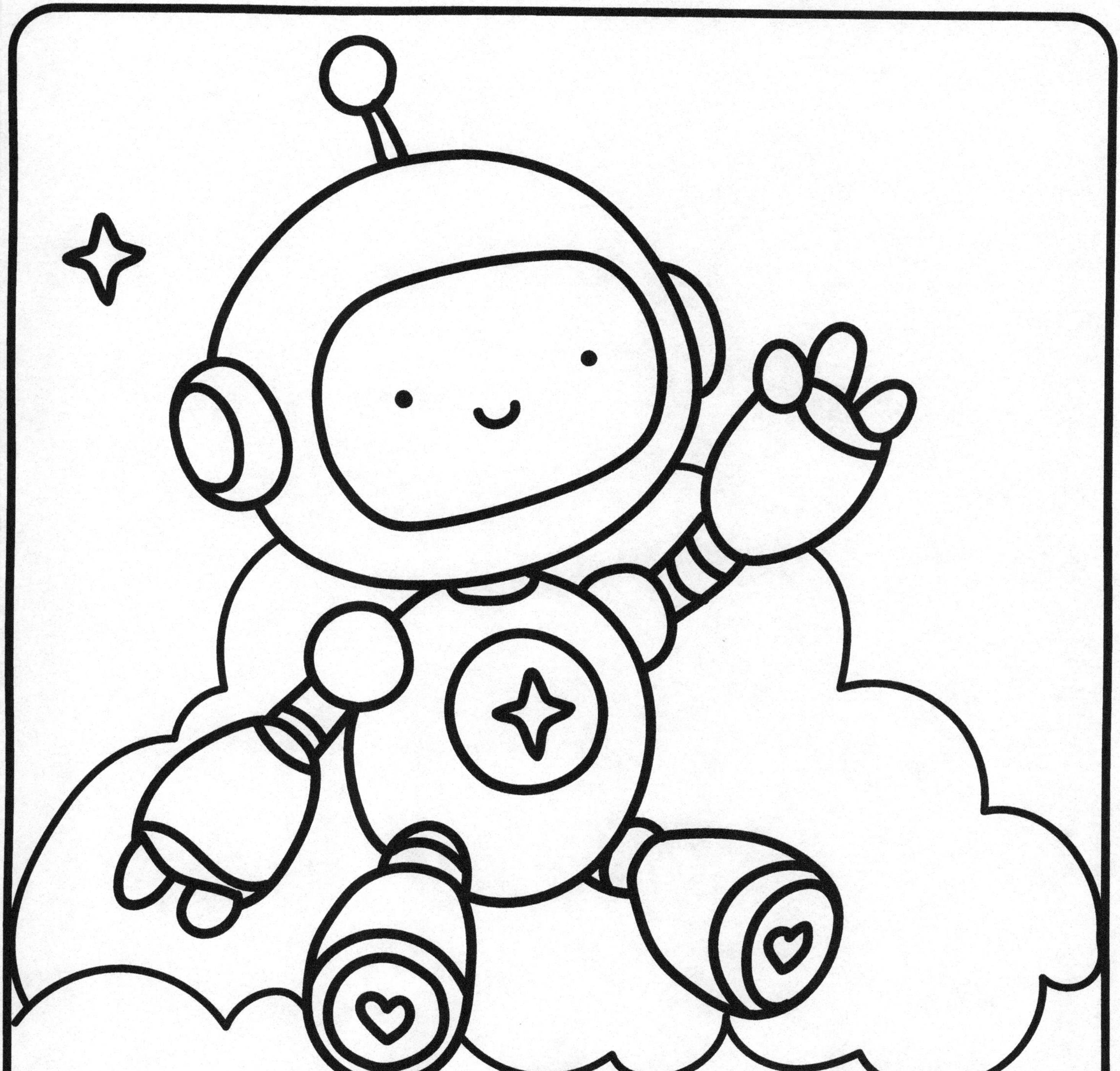

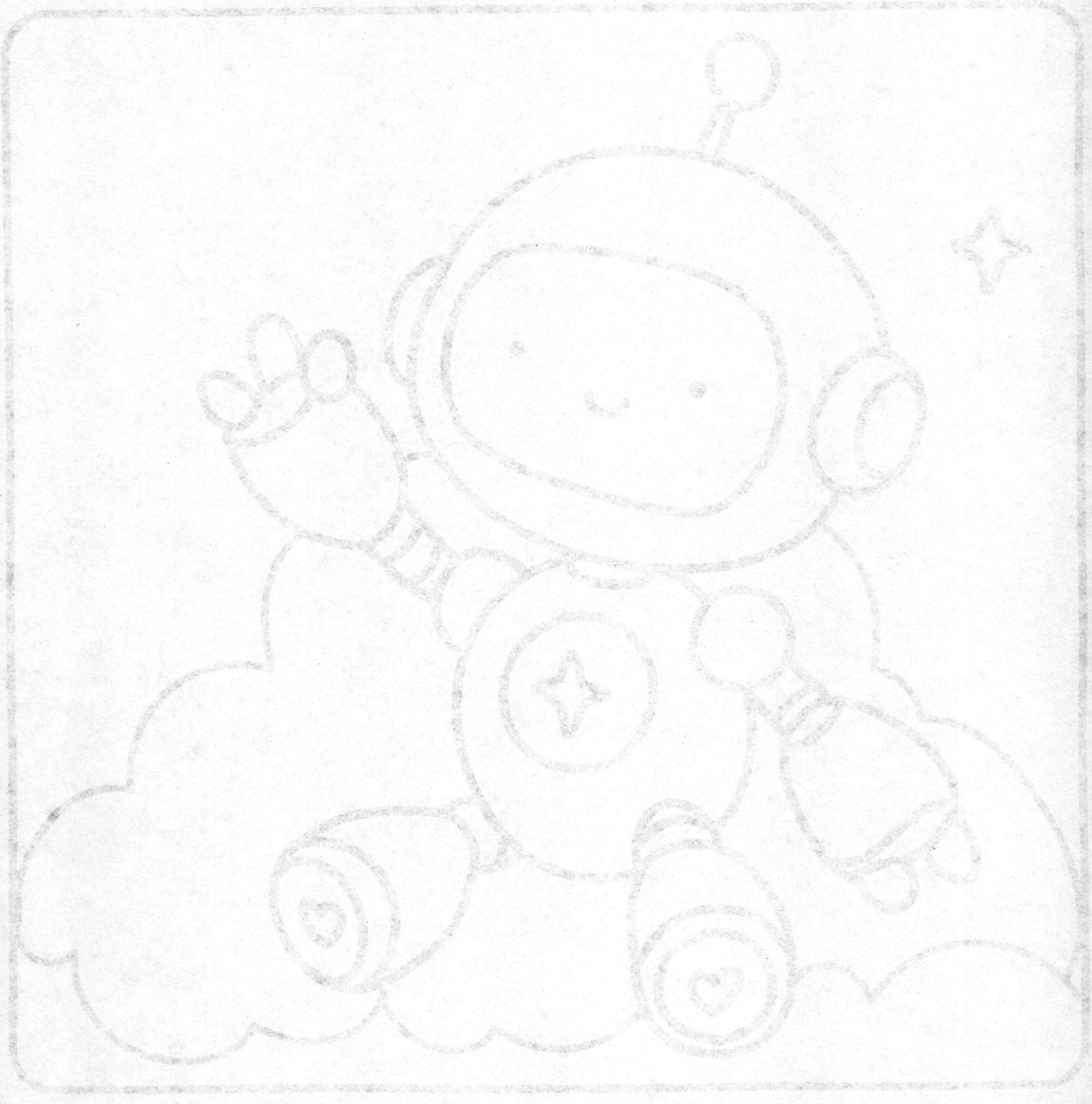

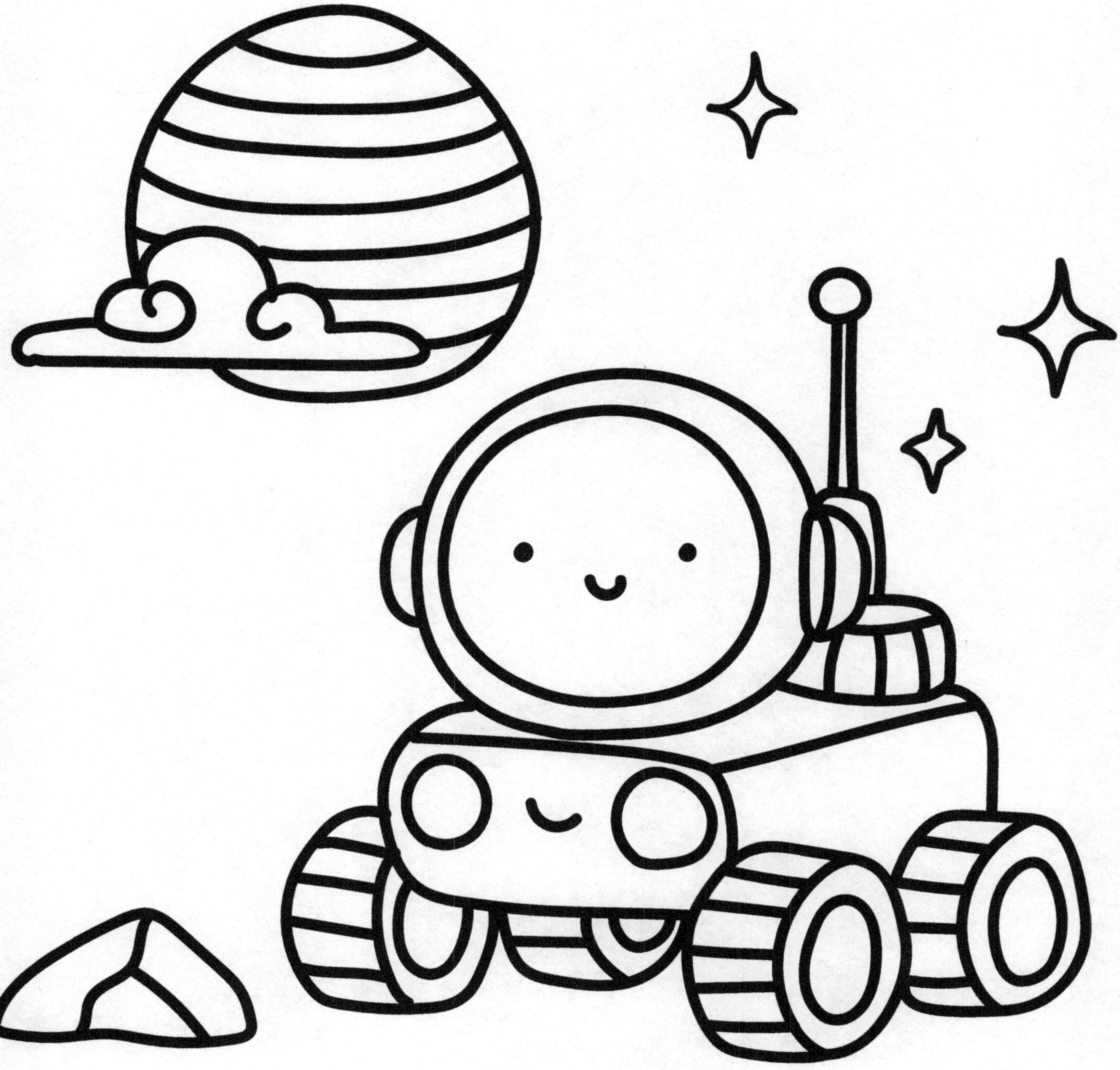

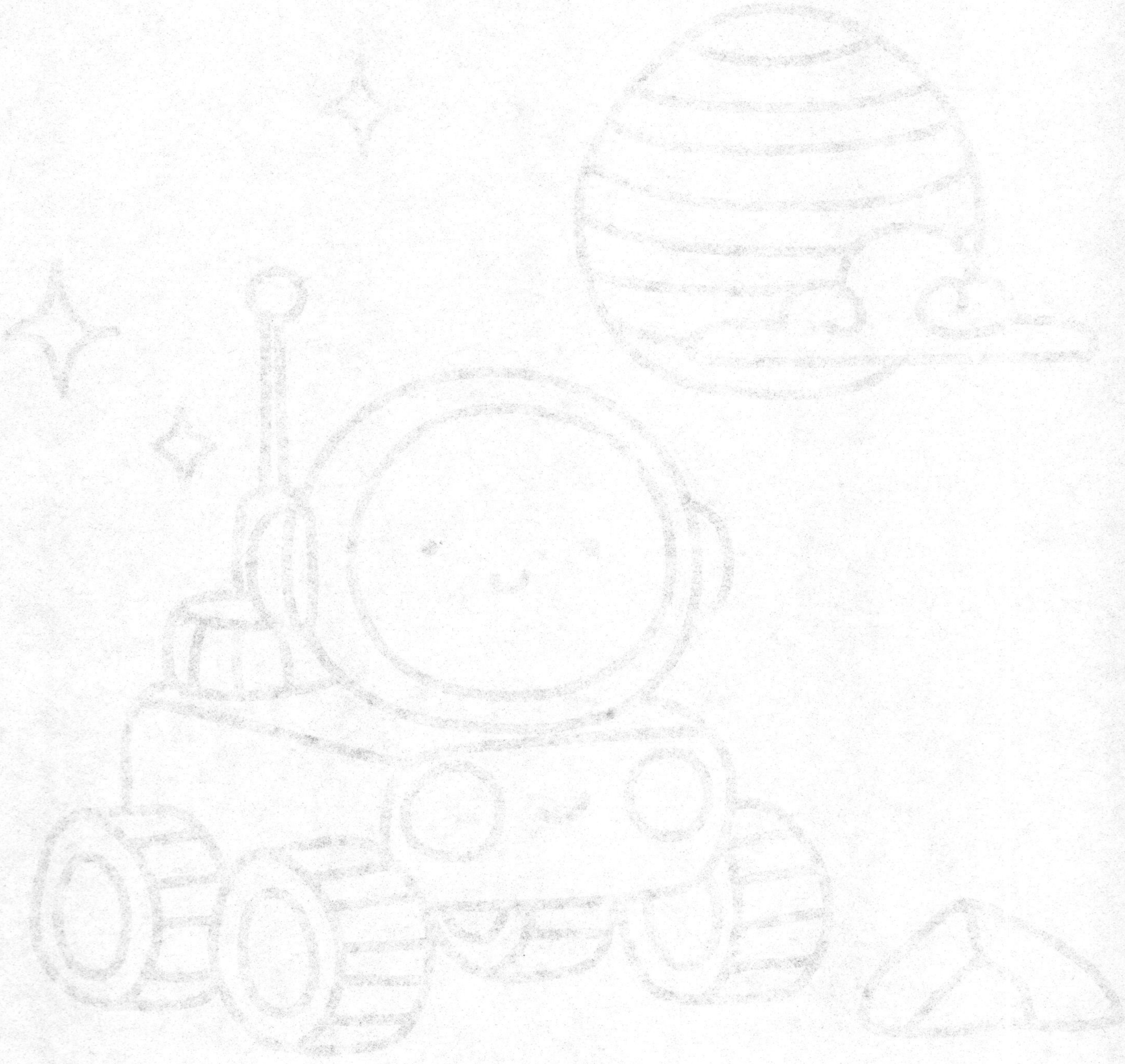

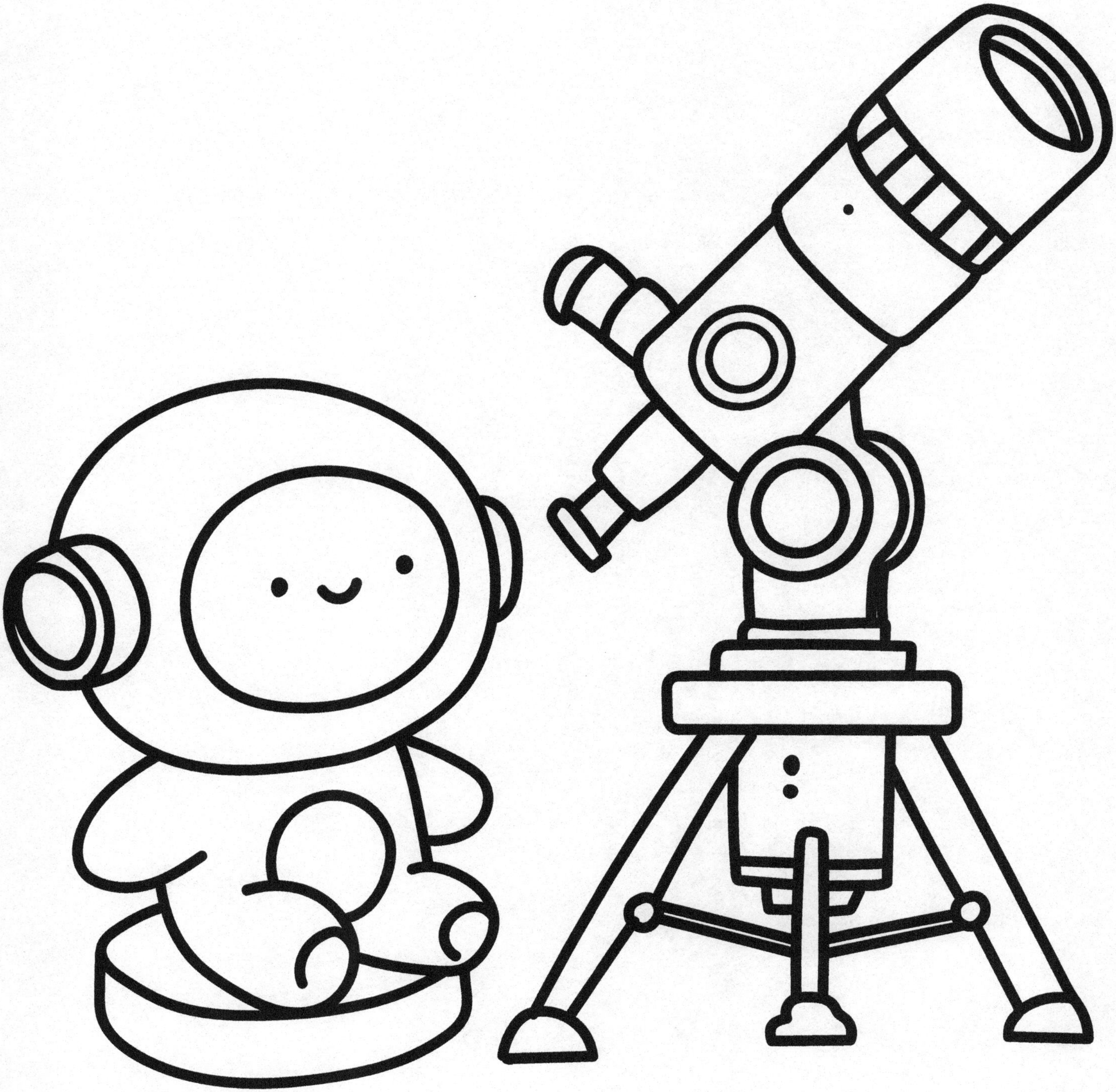

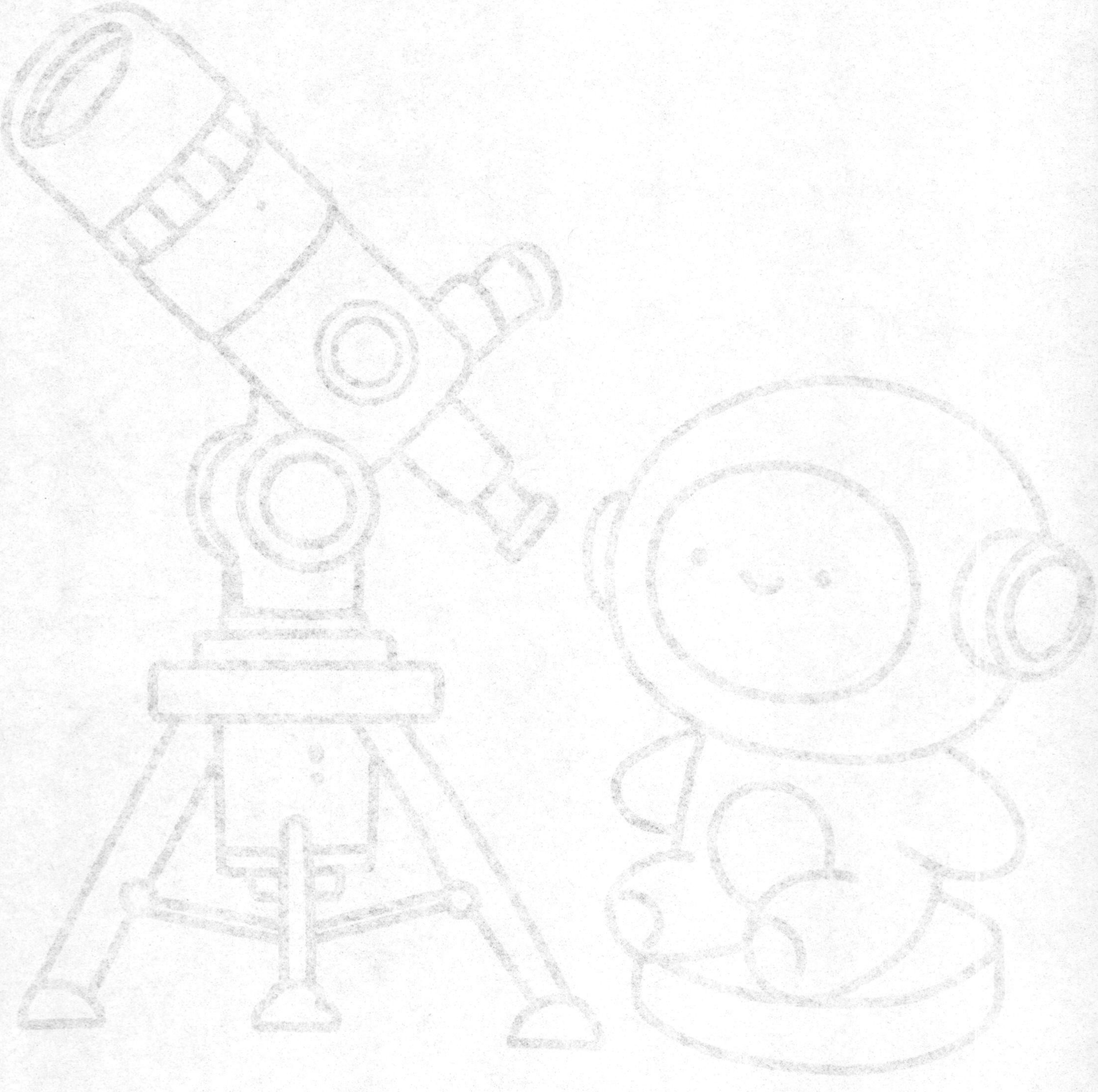

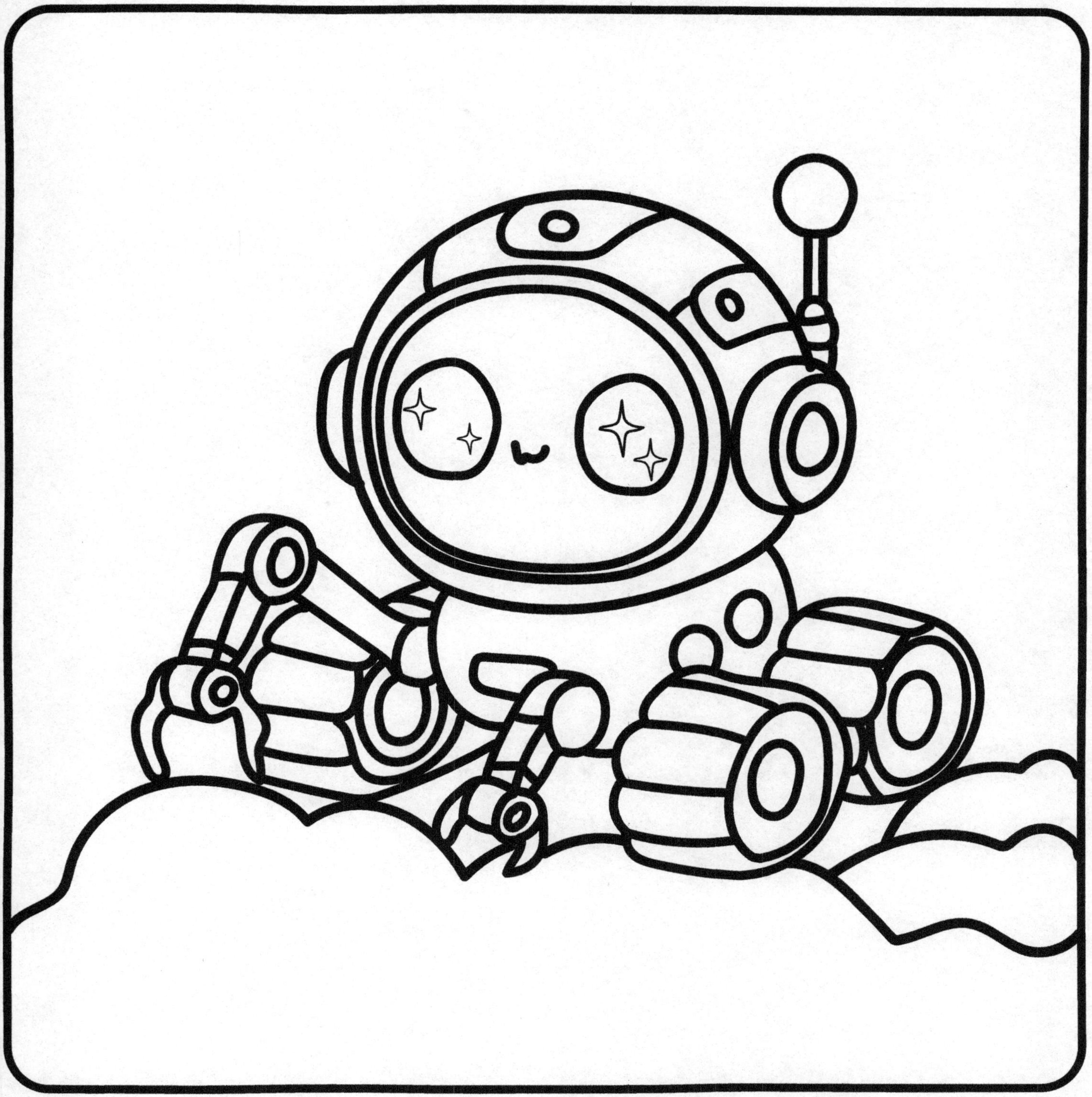

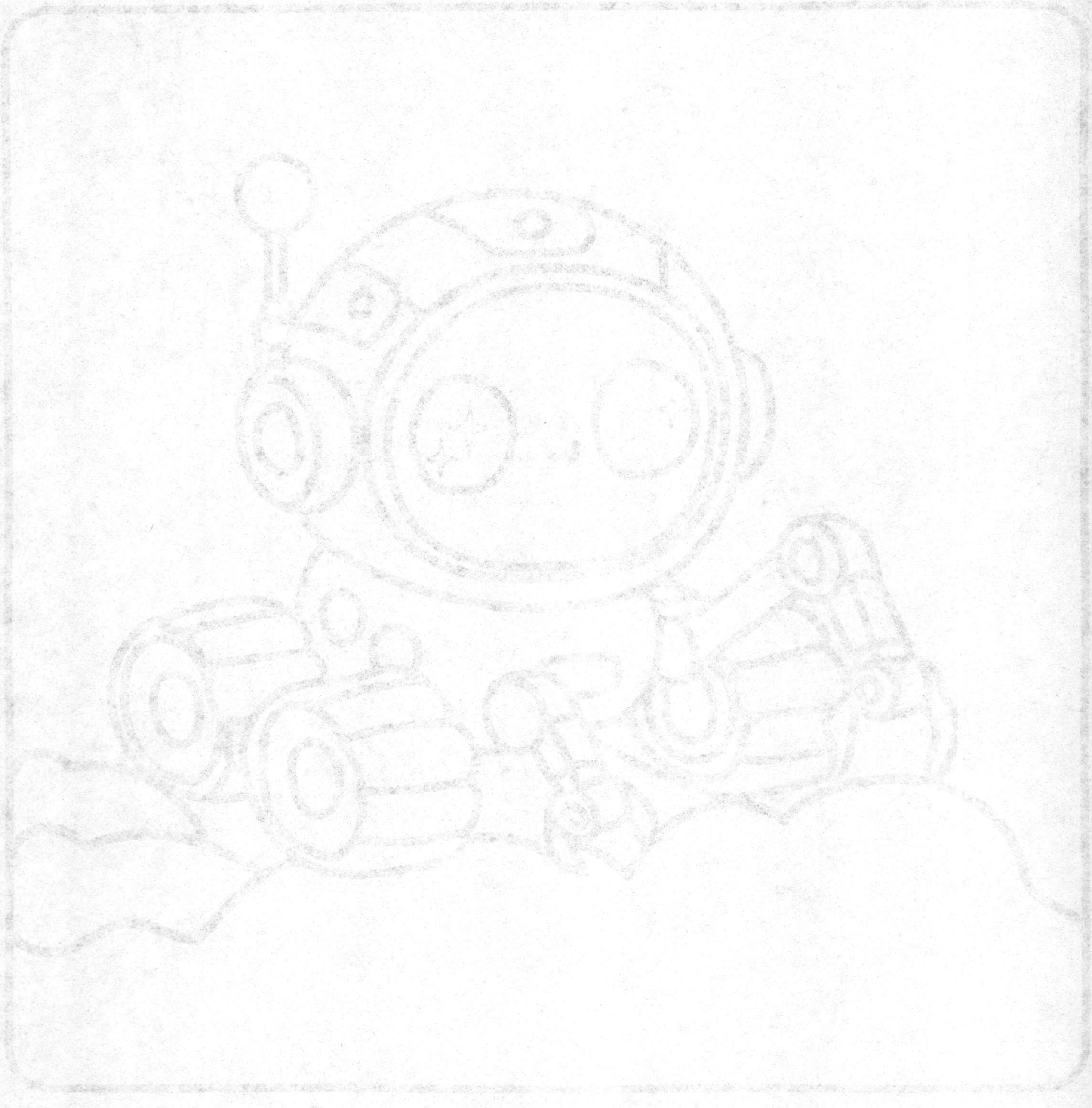

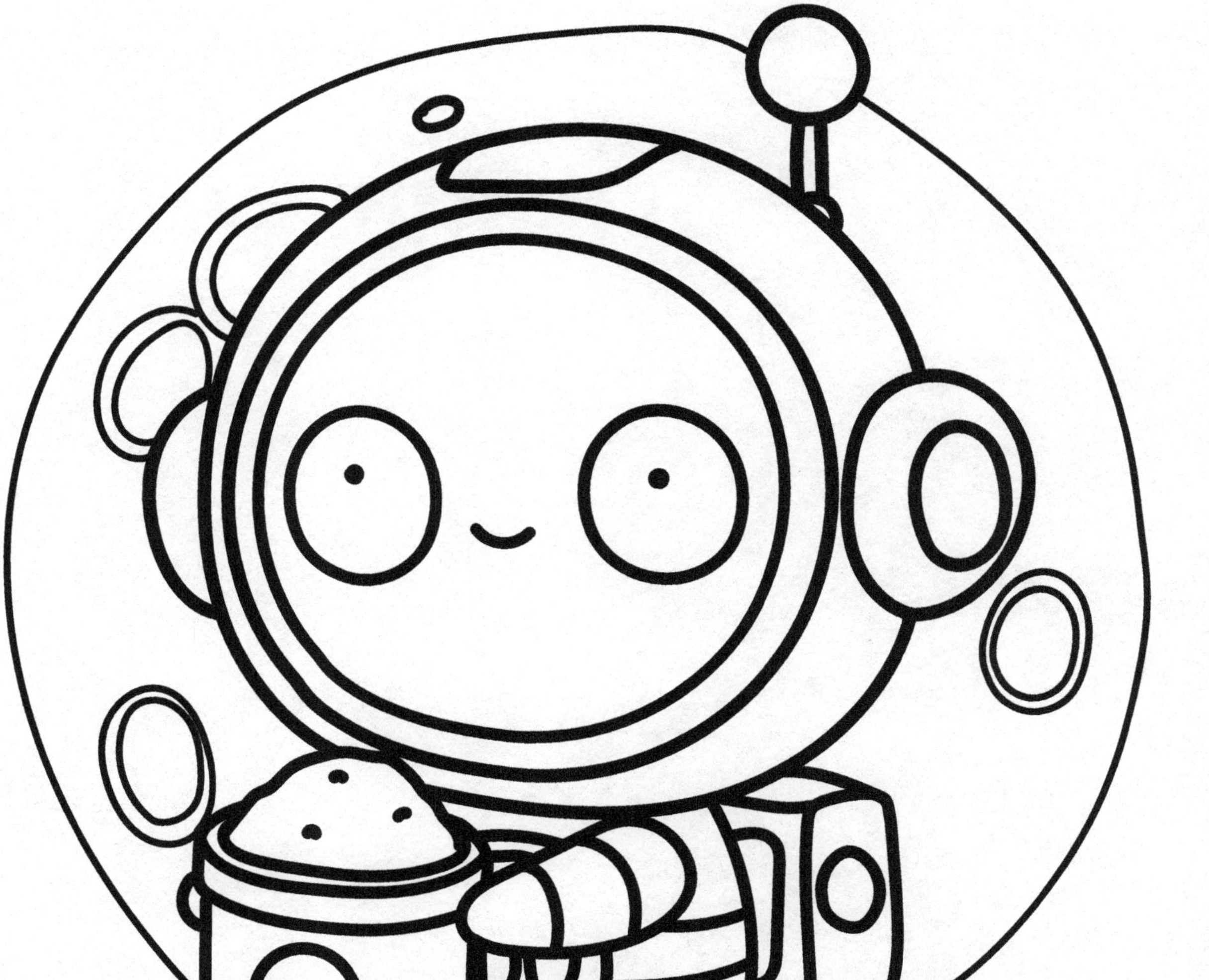

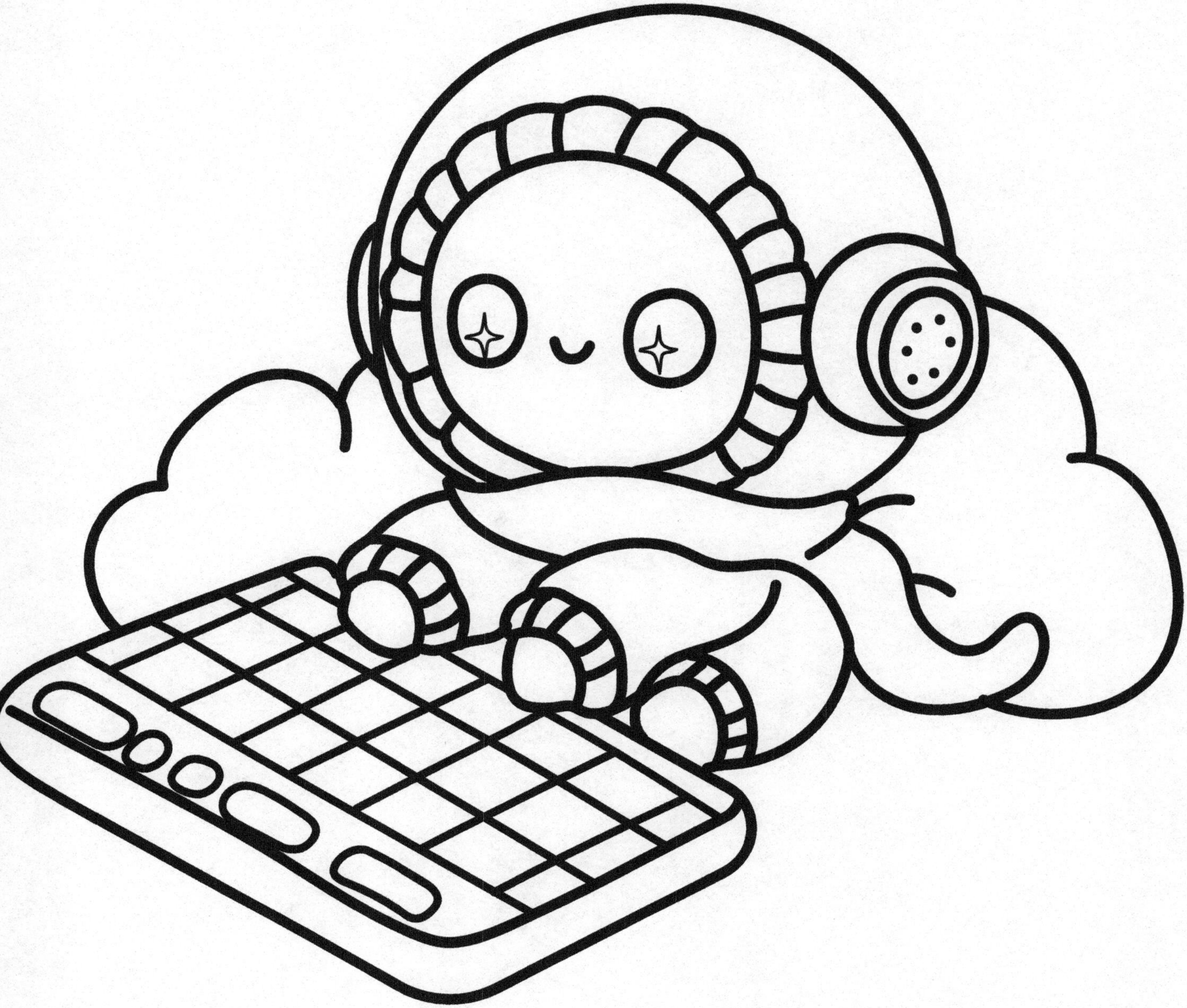

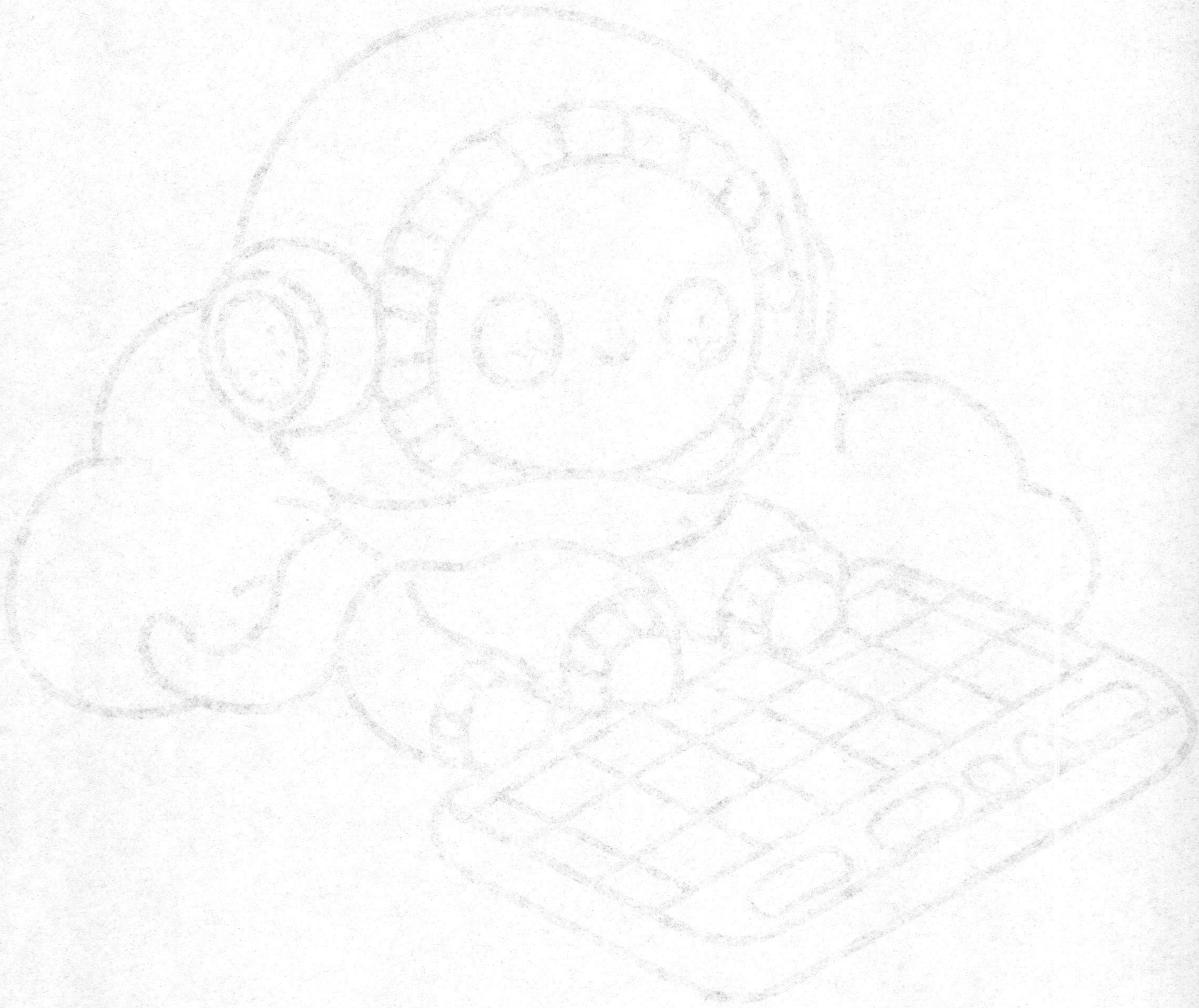

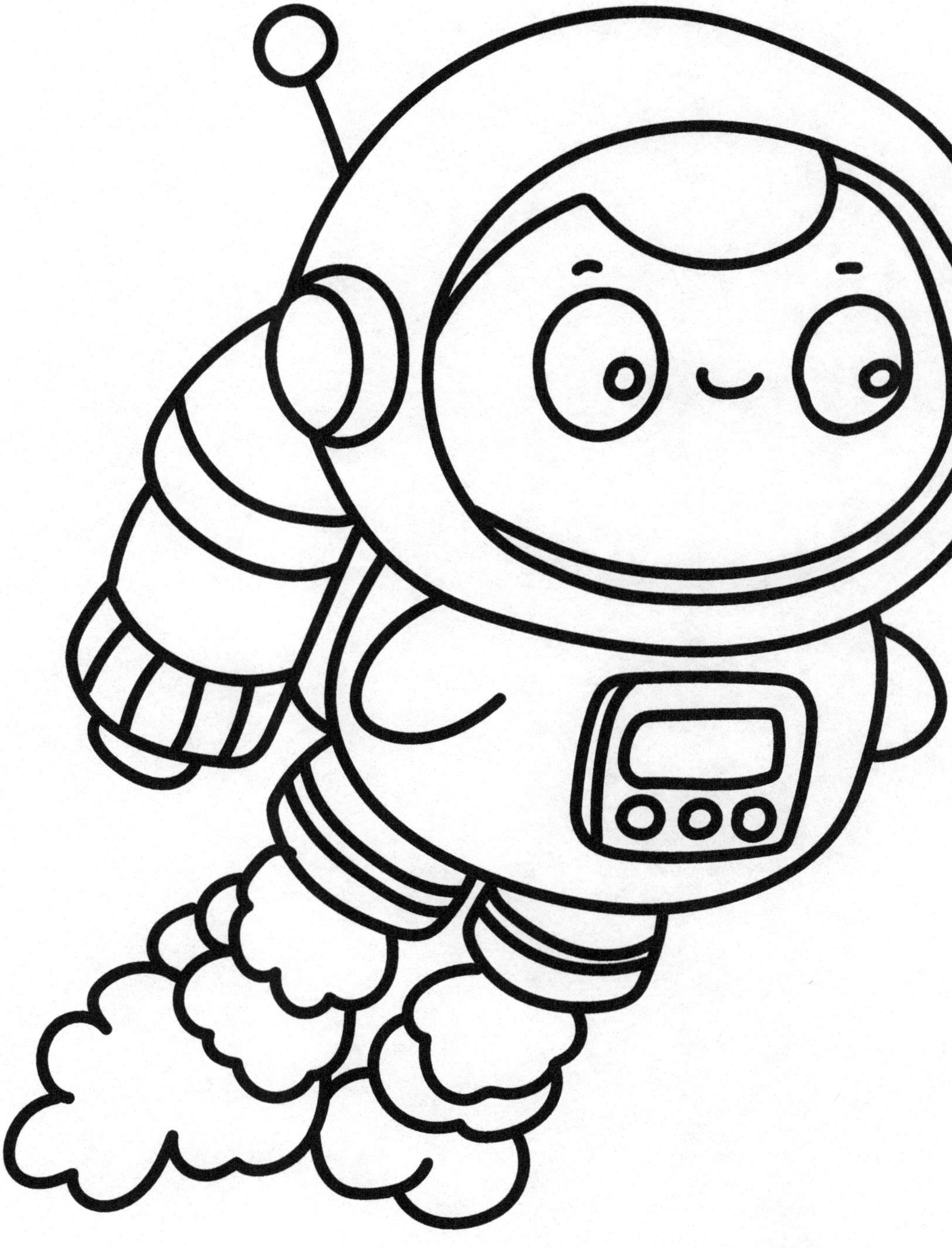

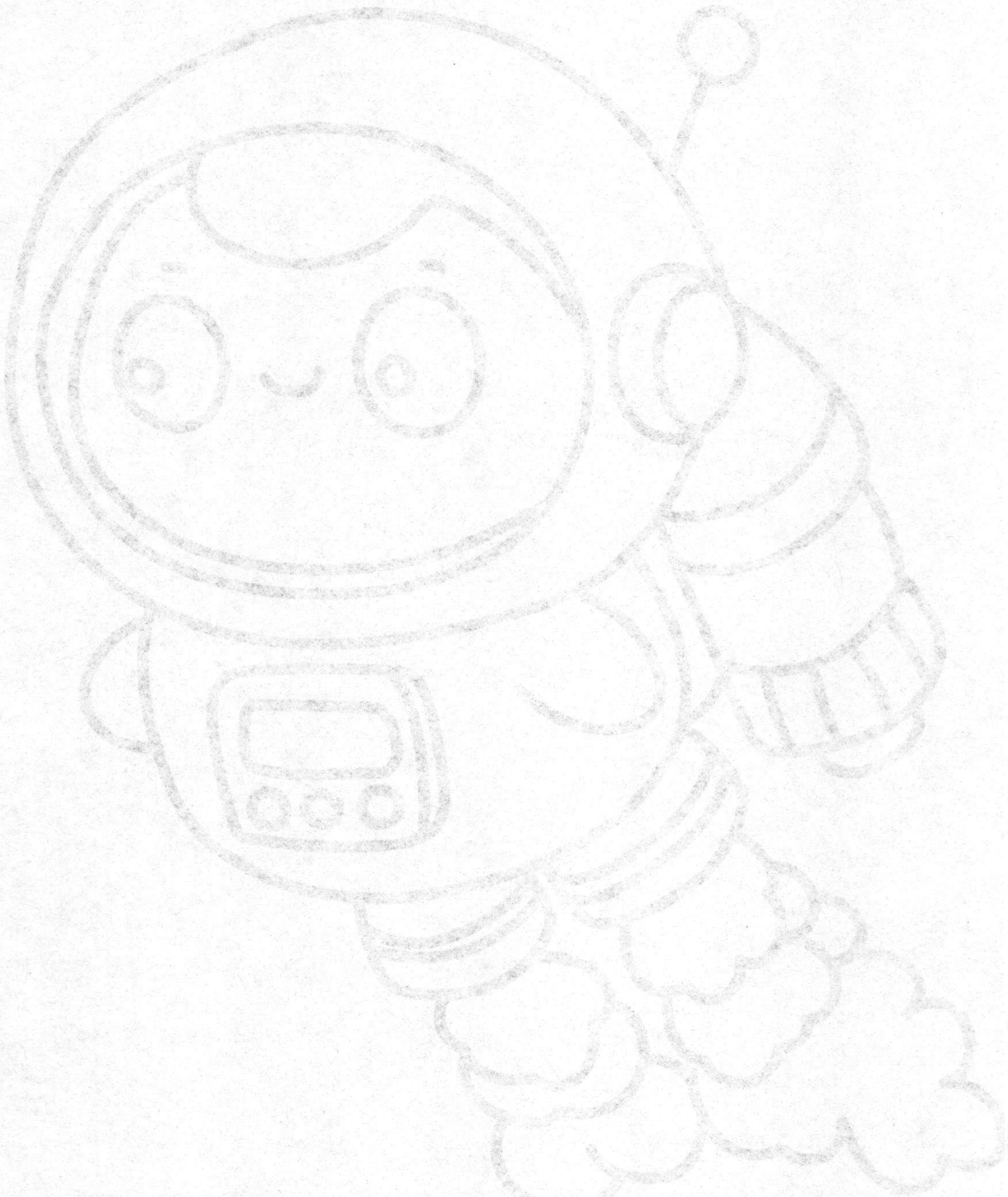

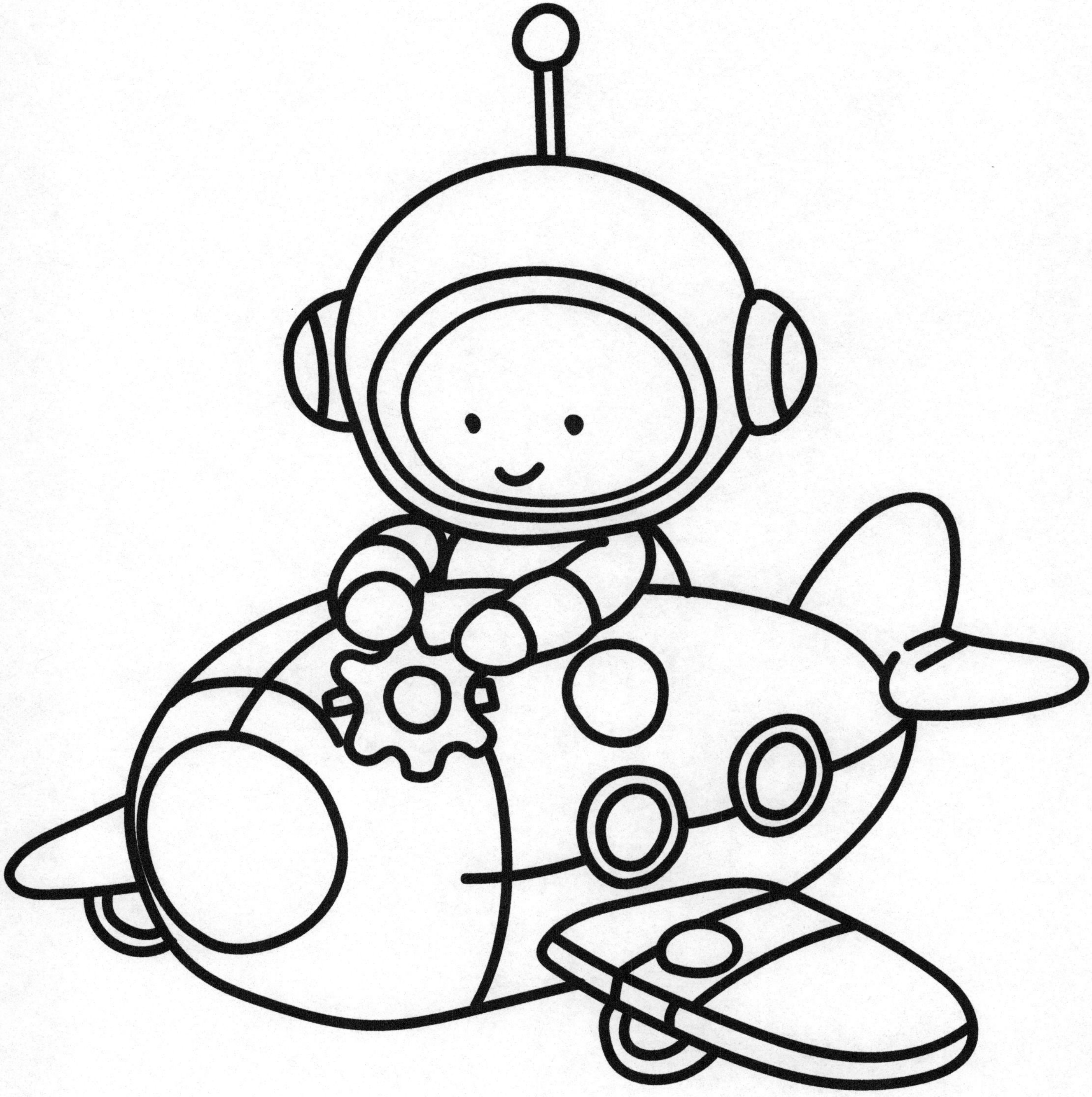

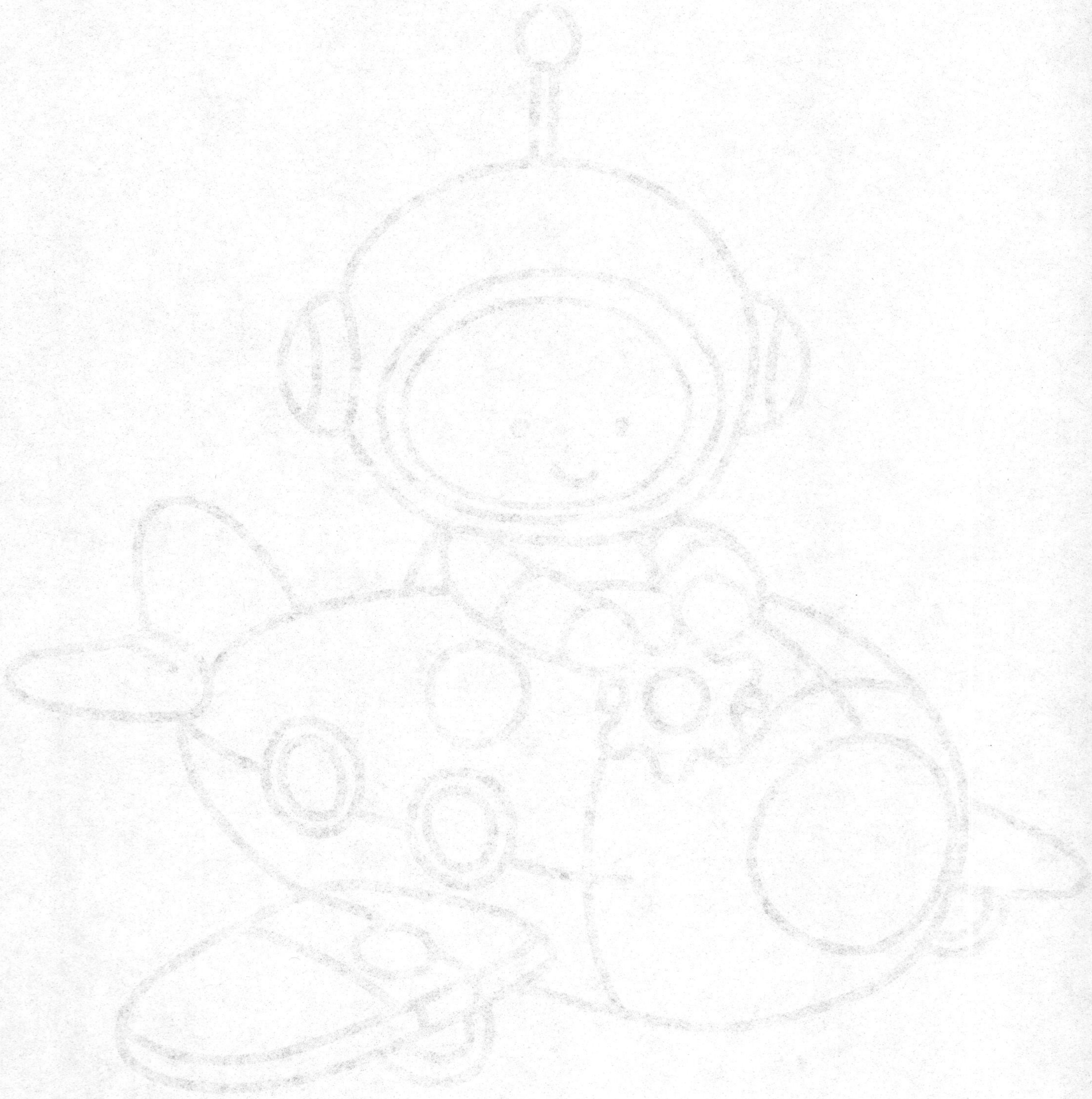

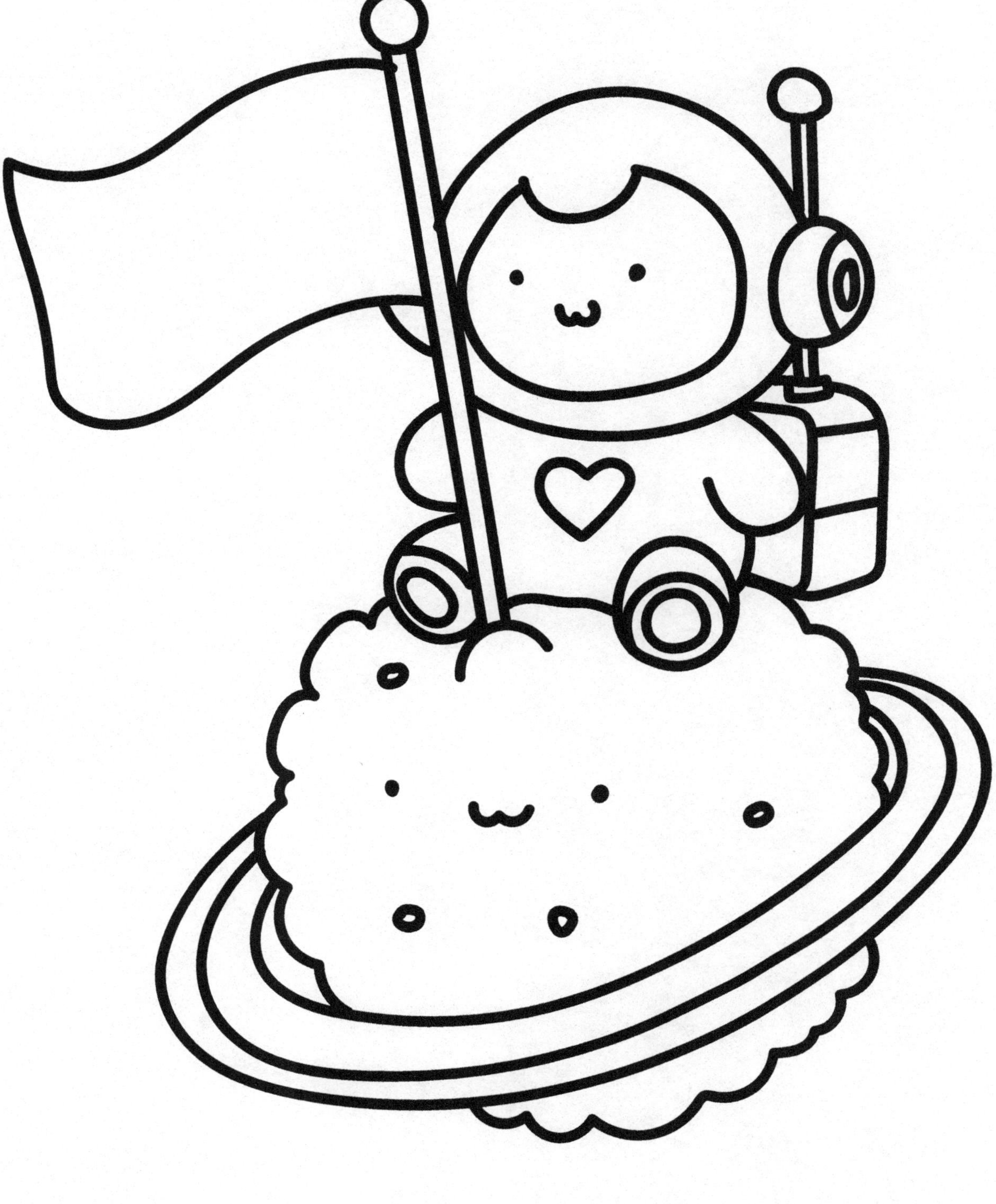

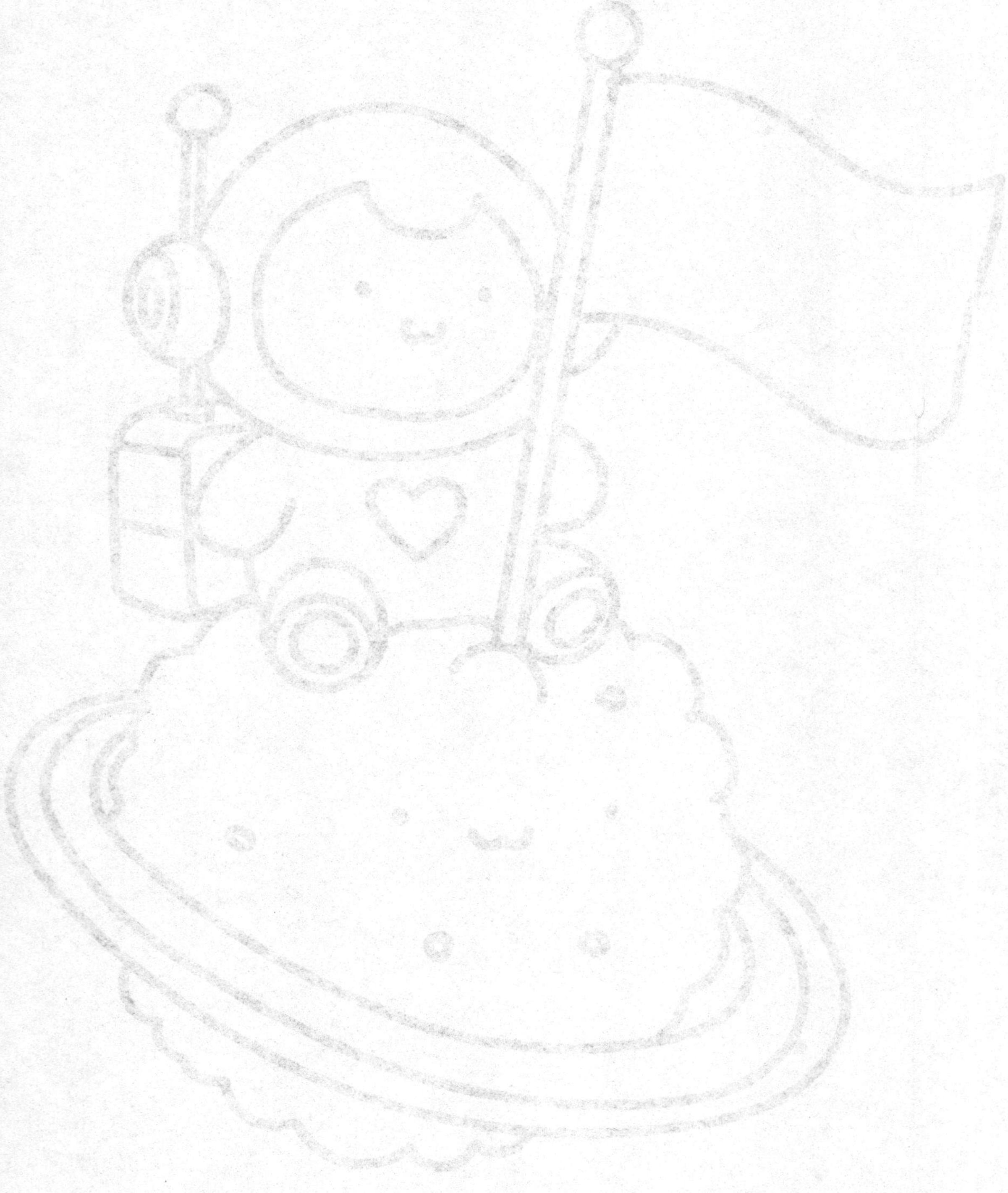

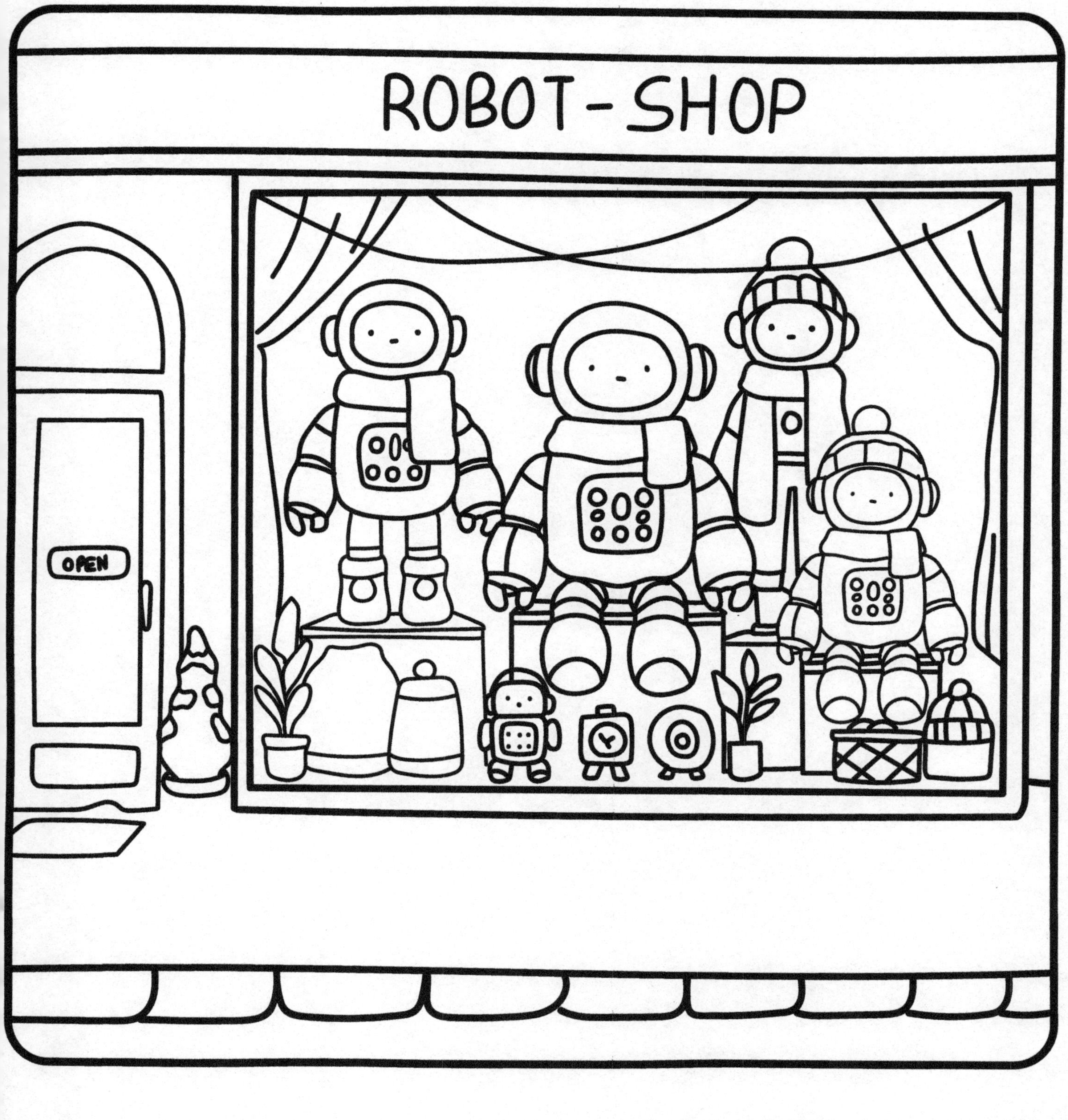
ROBOT - SHOP
OPEN

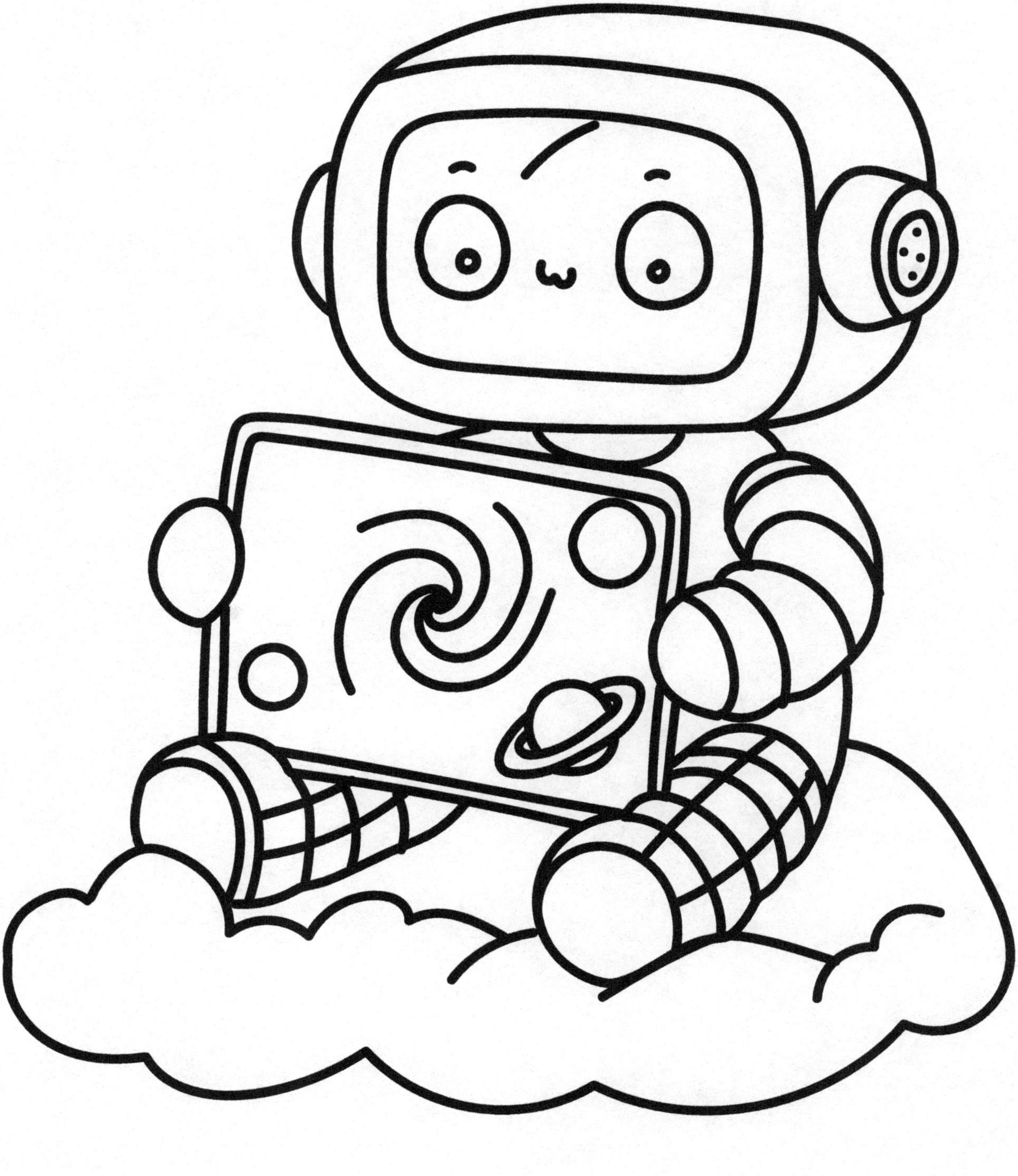

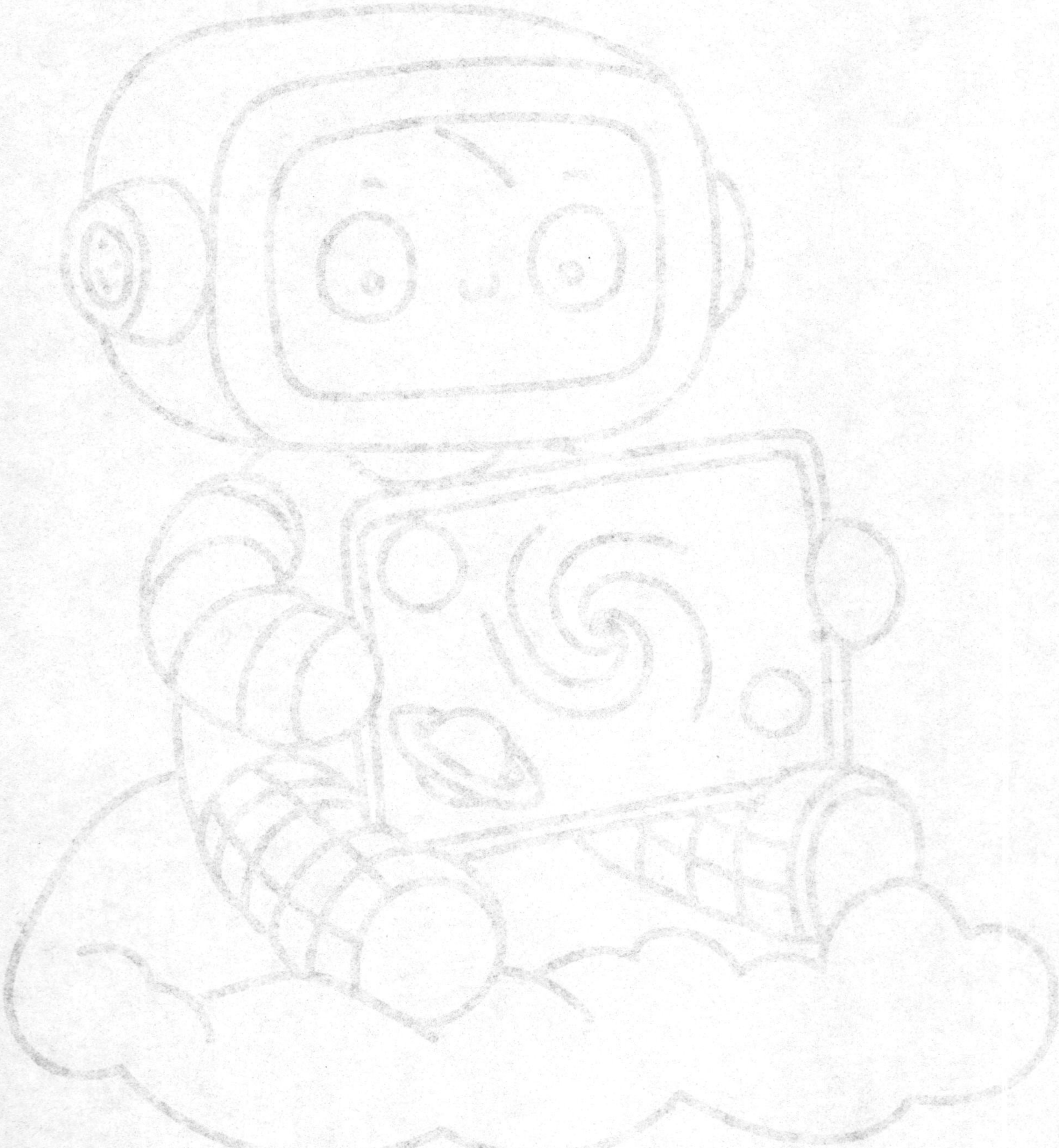

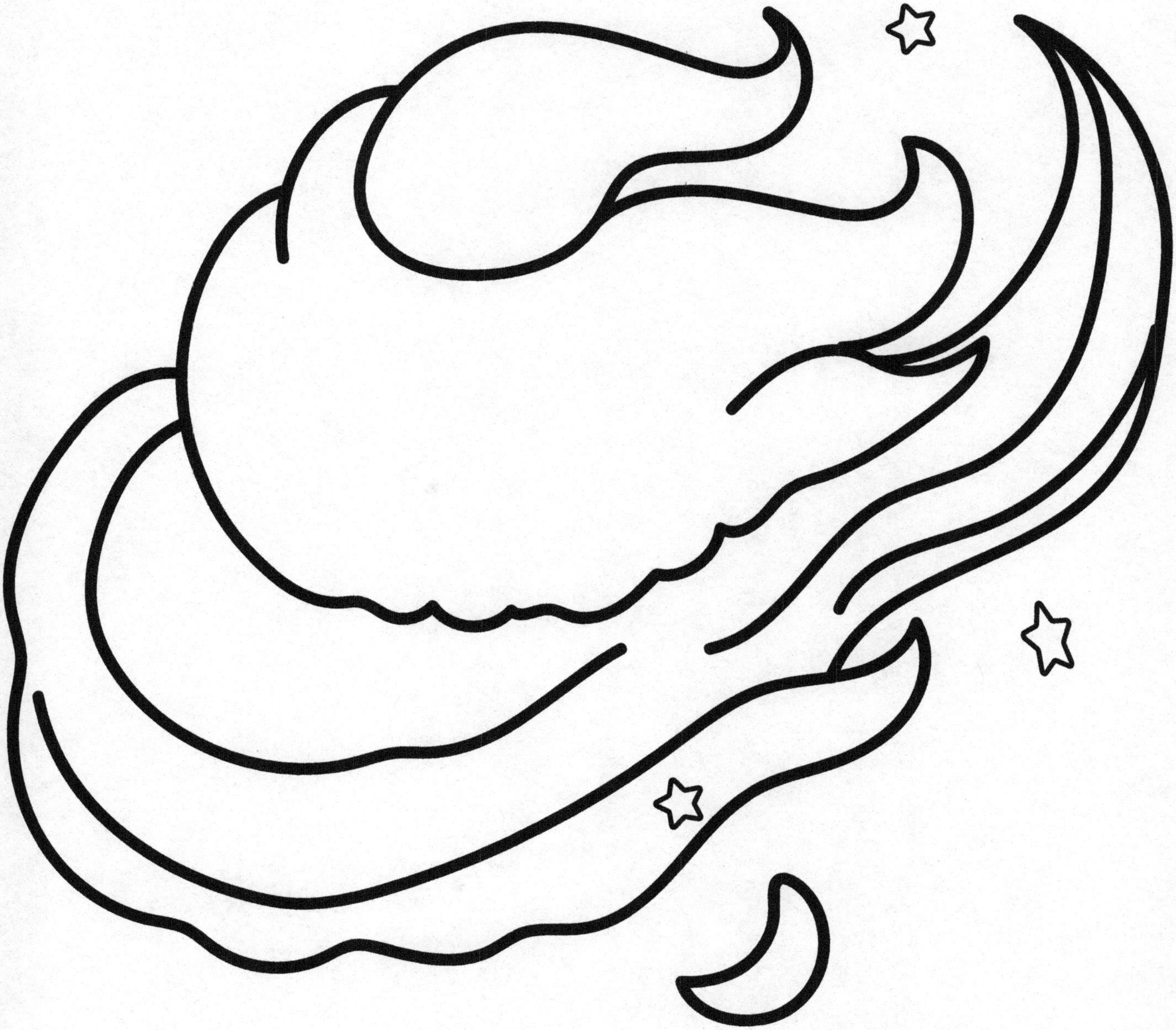

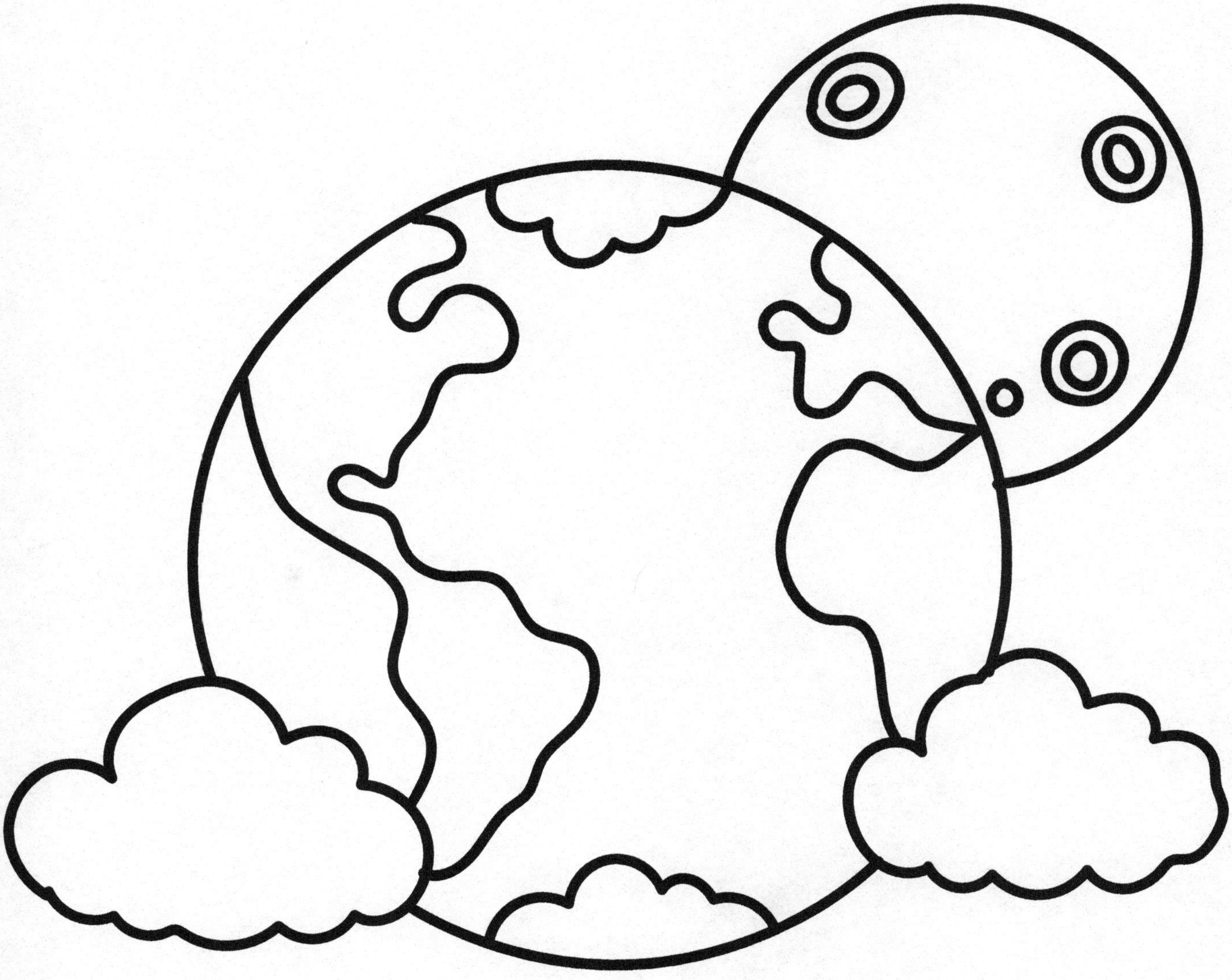

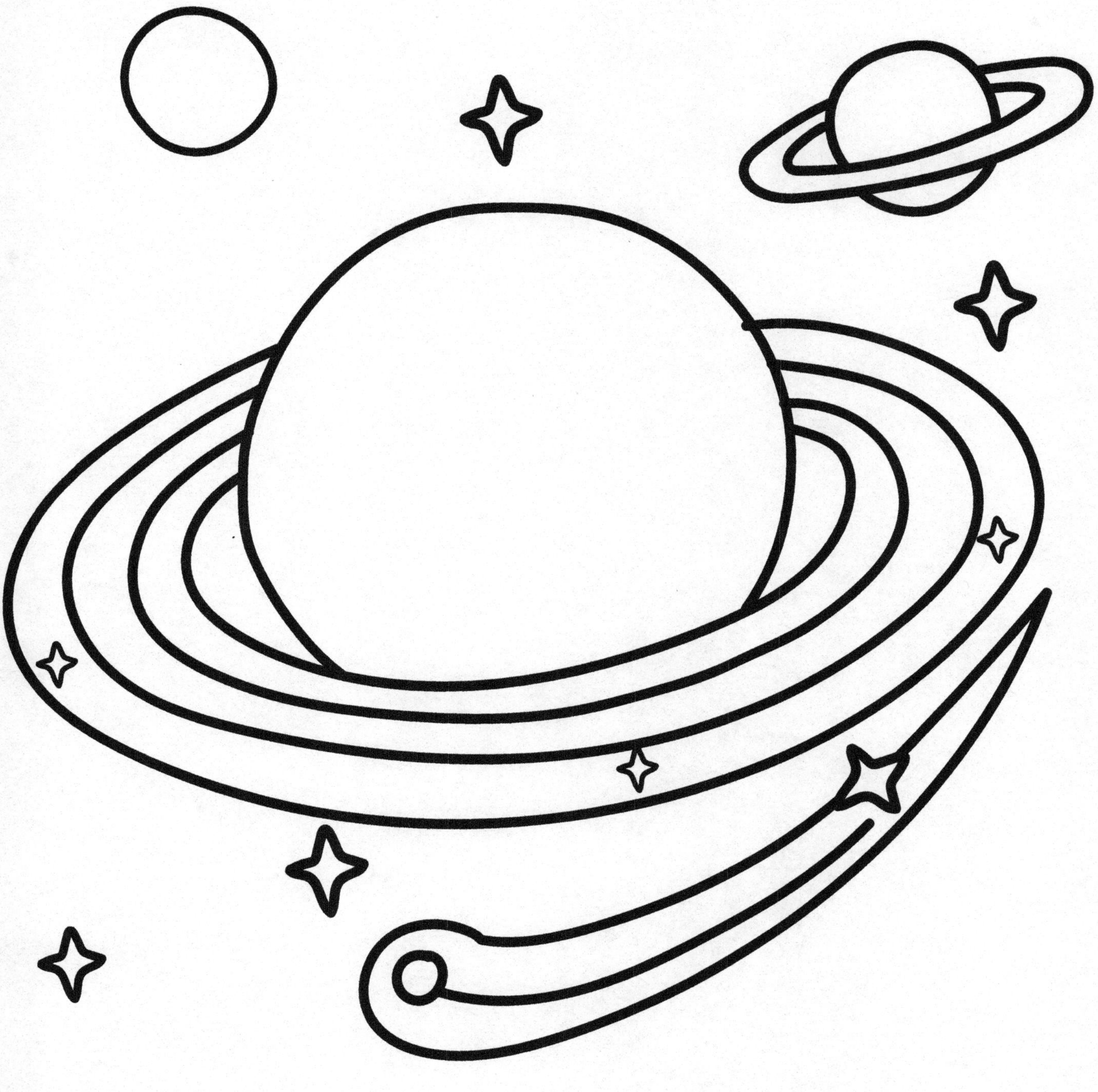

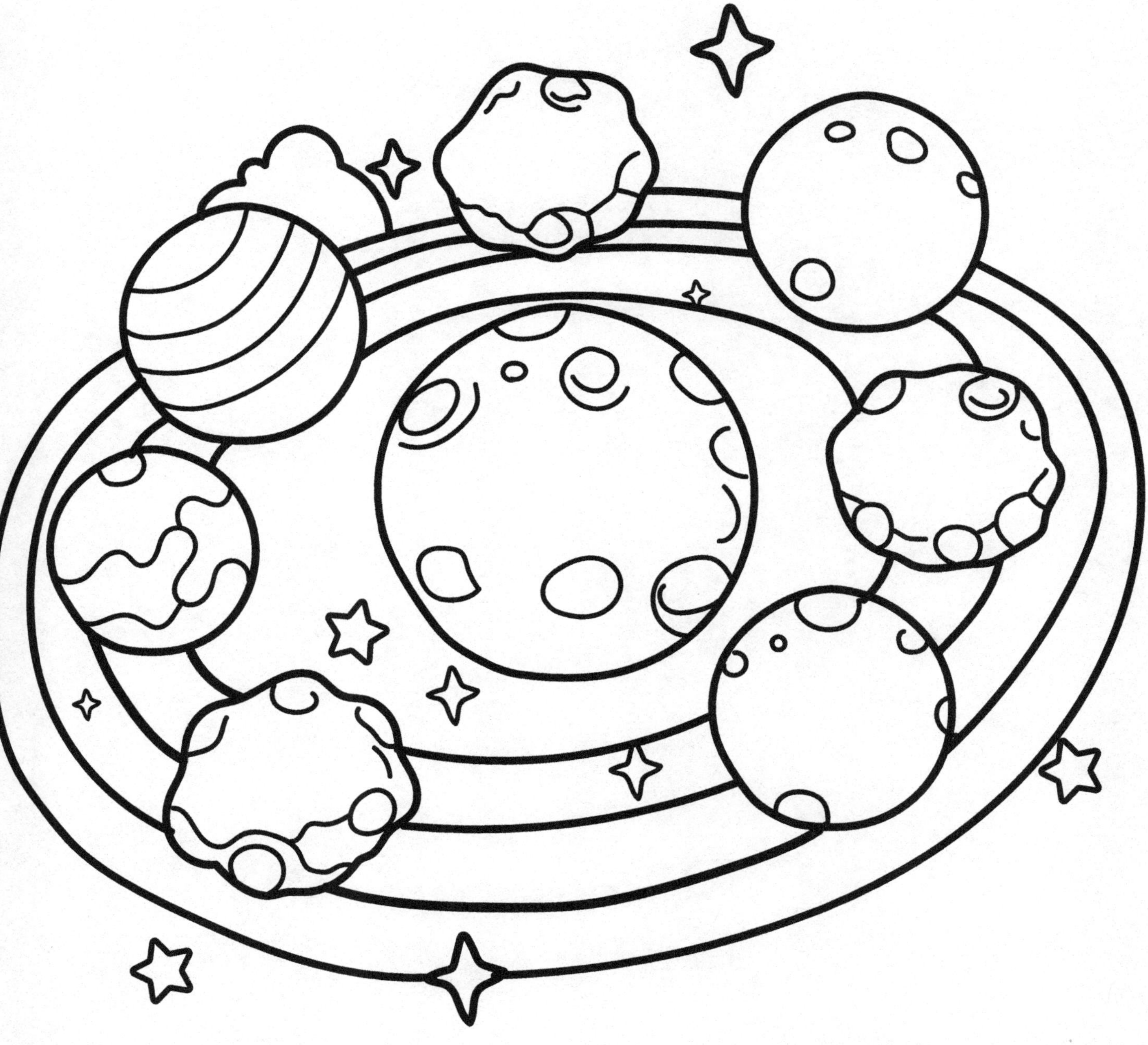

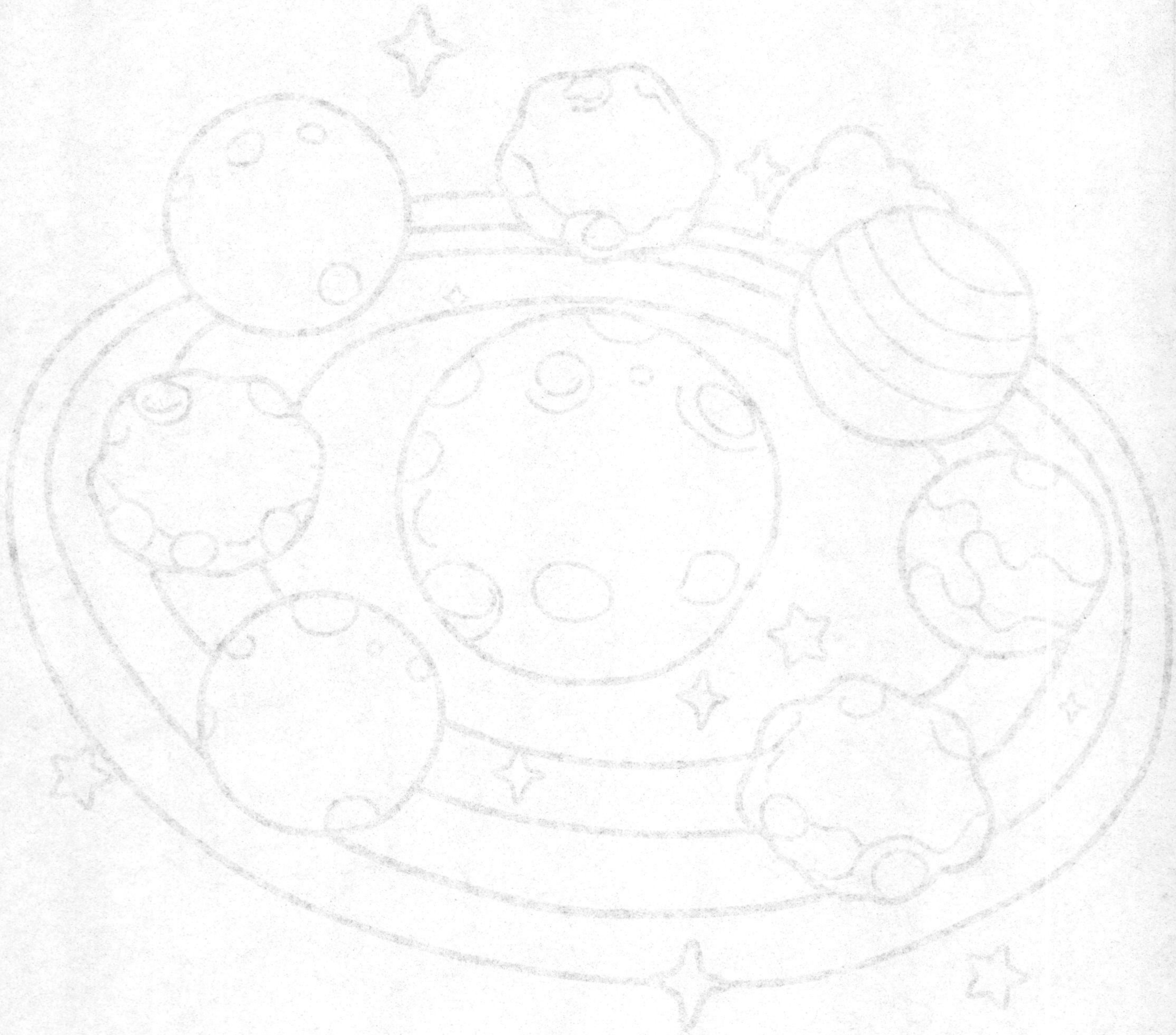

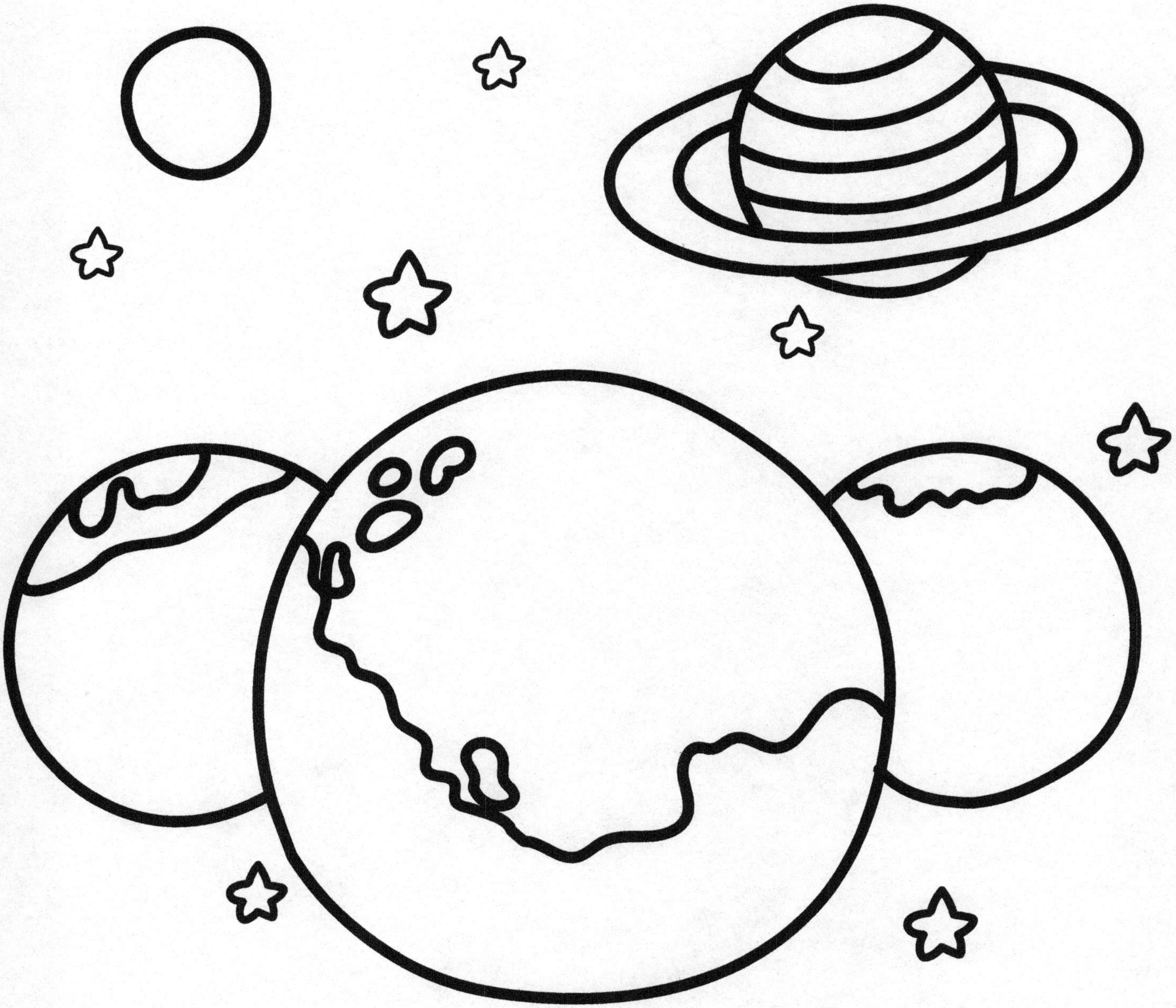

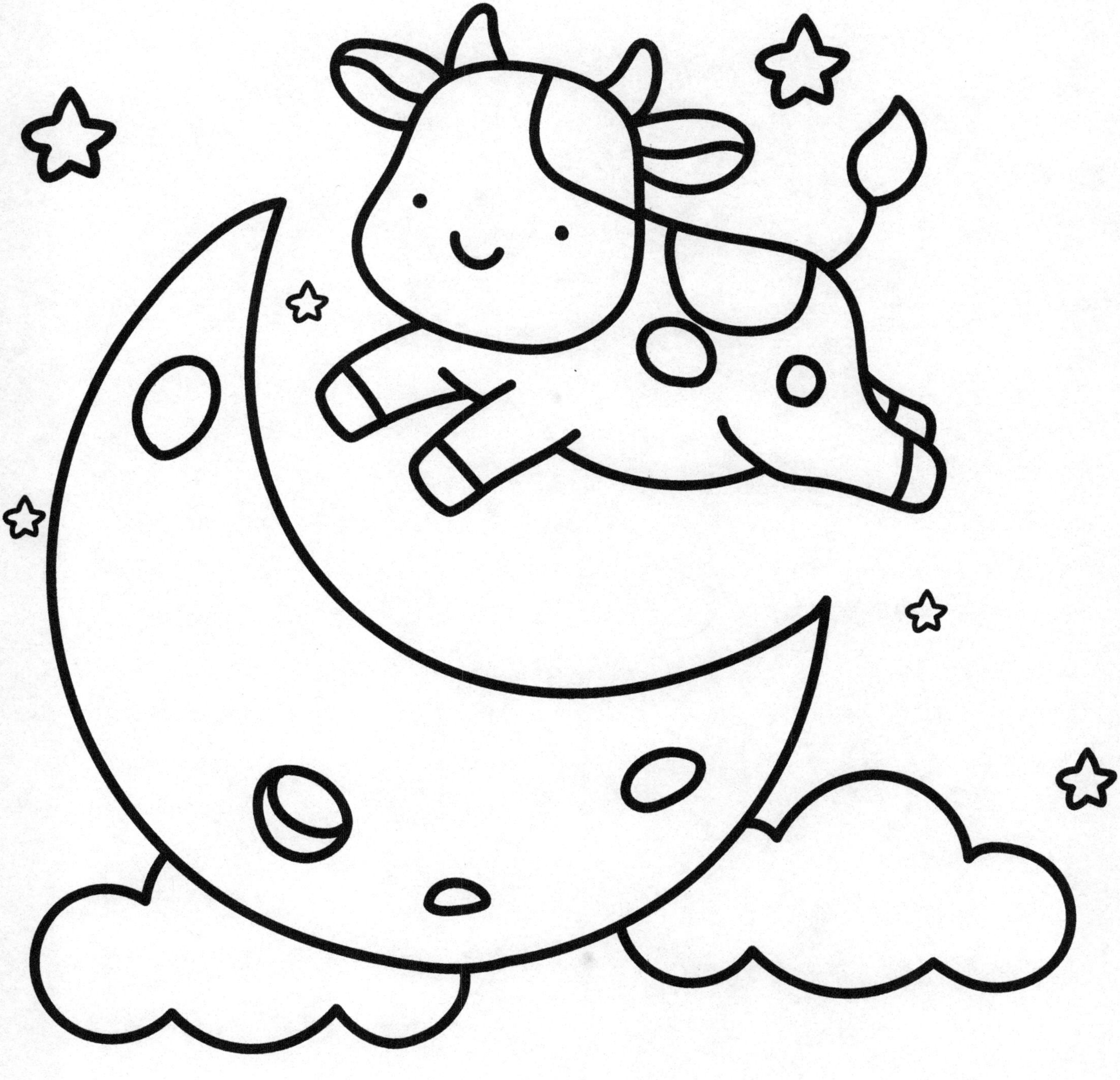

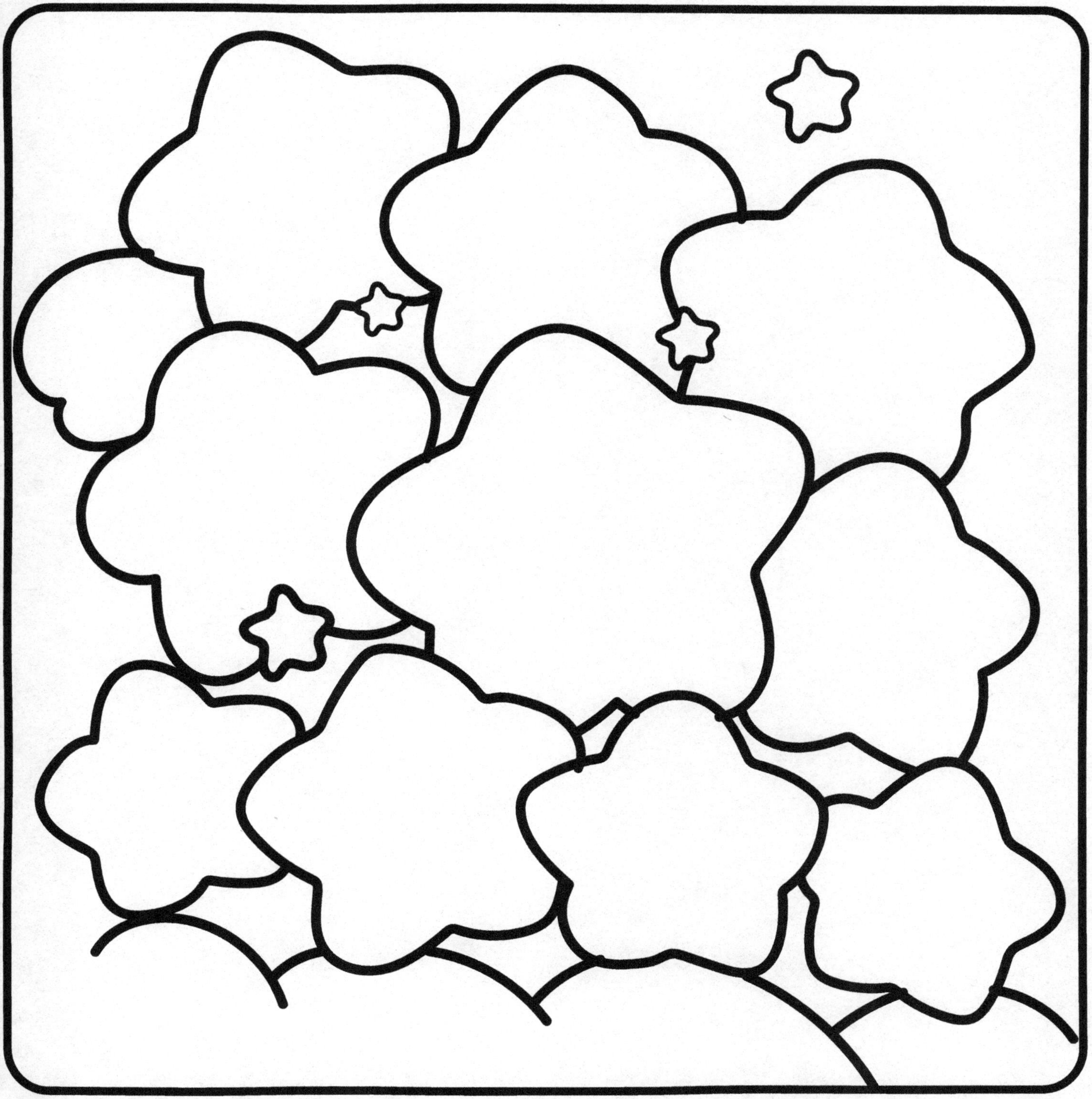

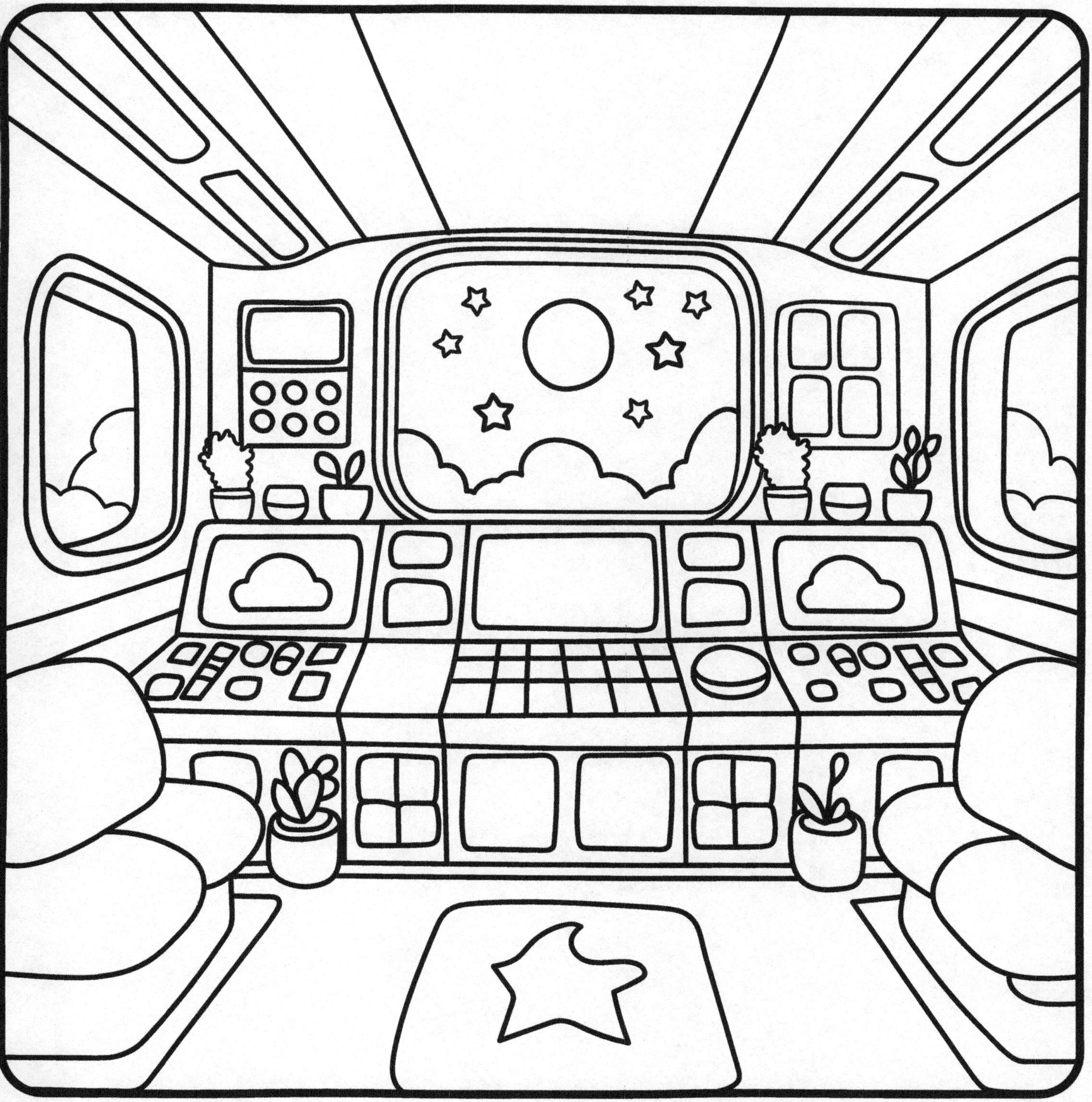

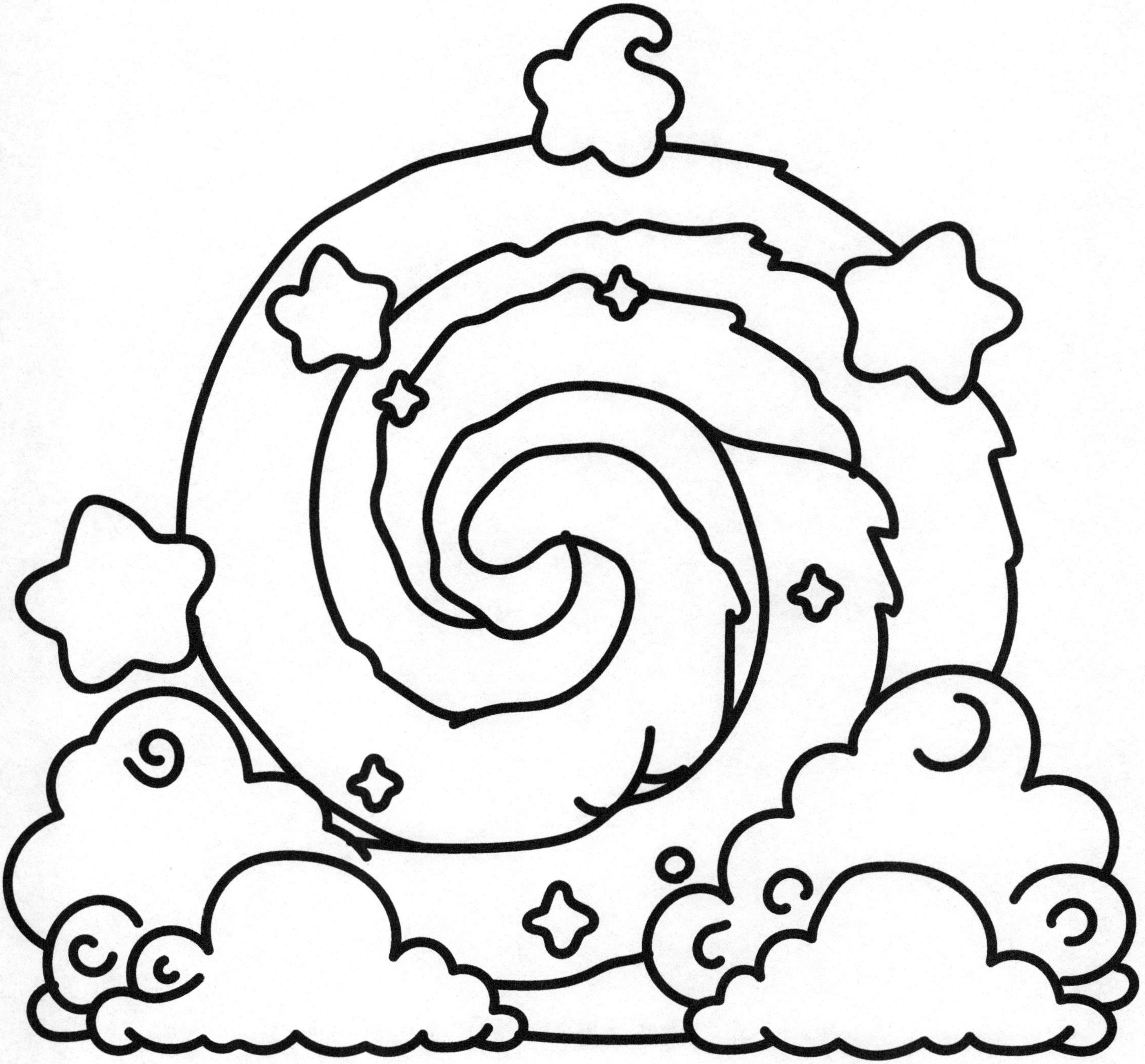

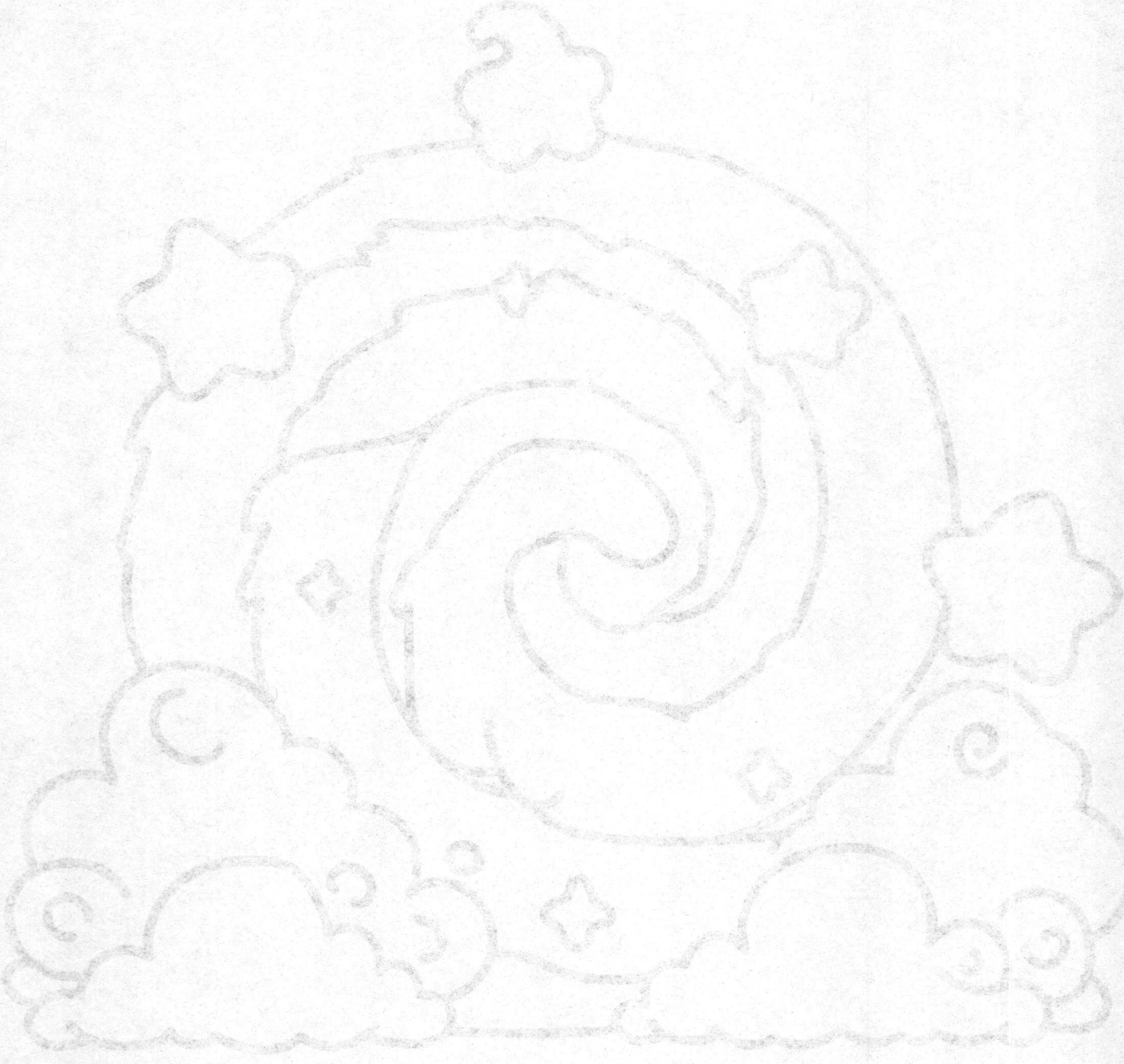

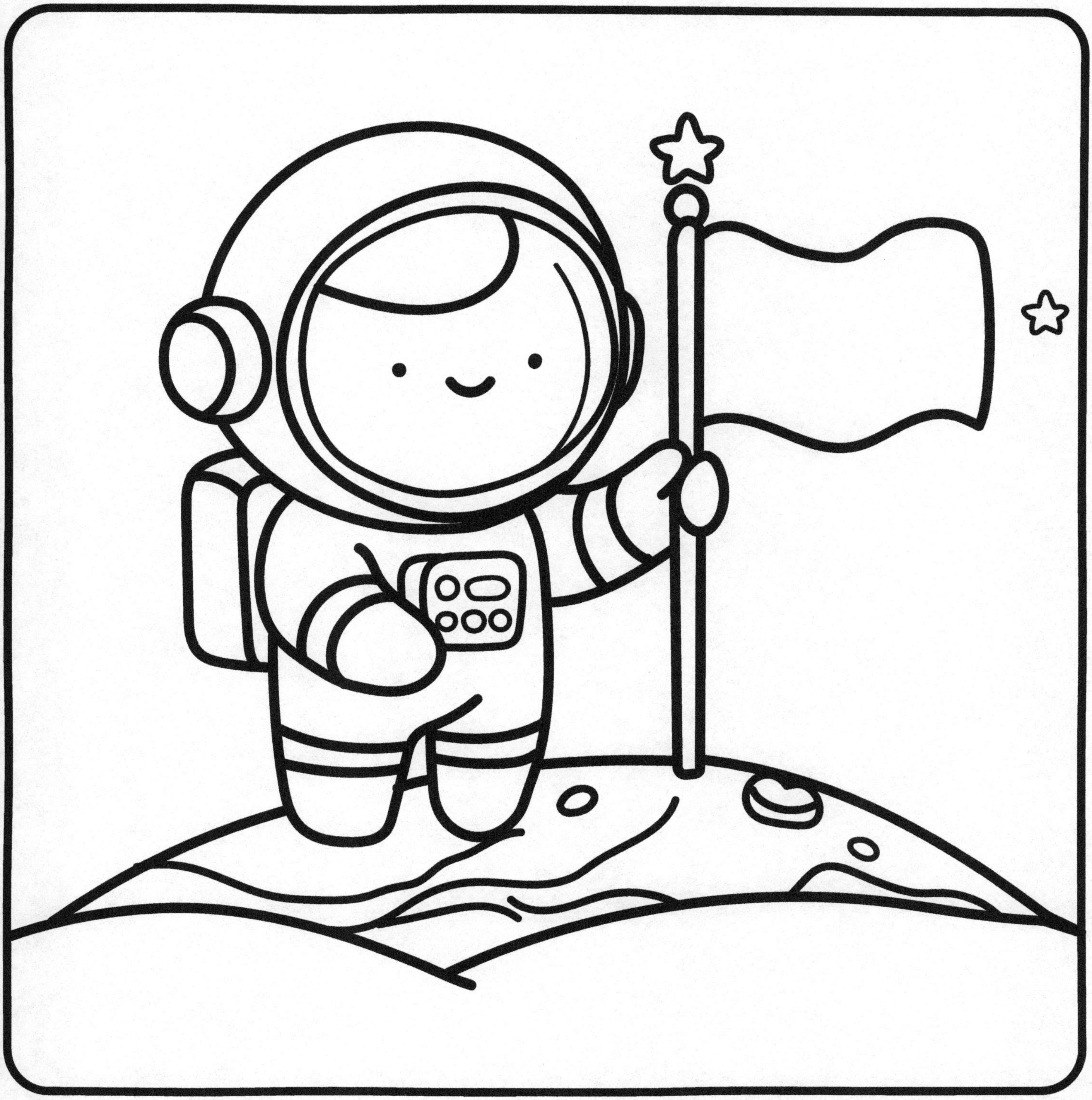

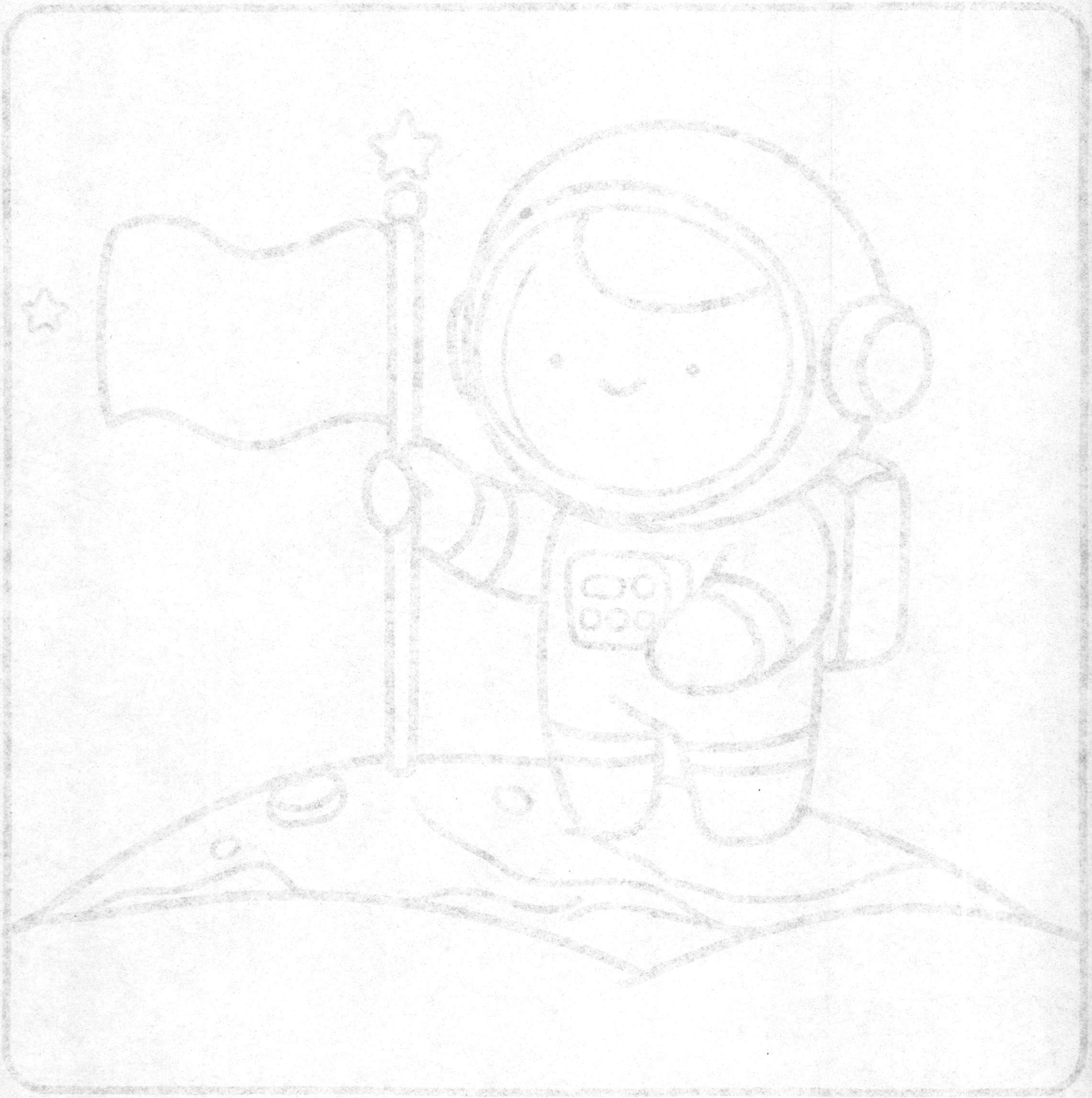

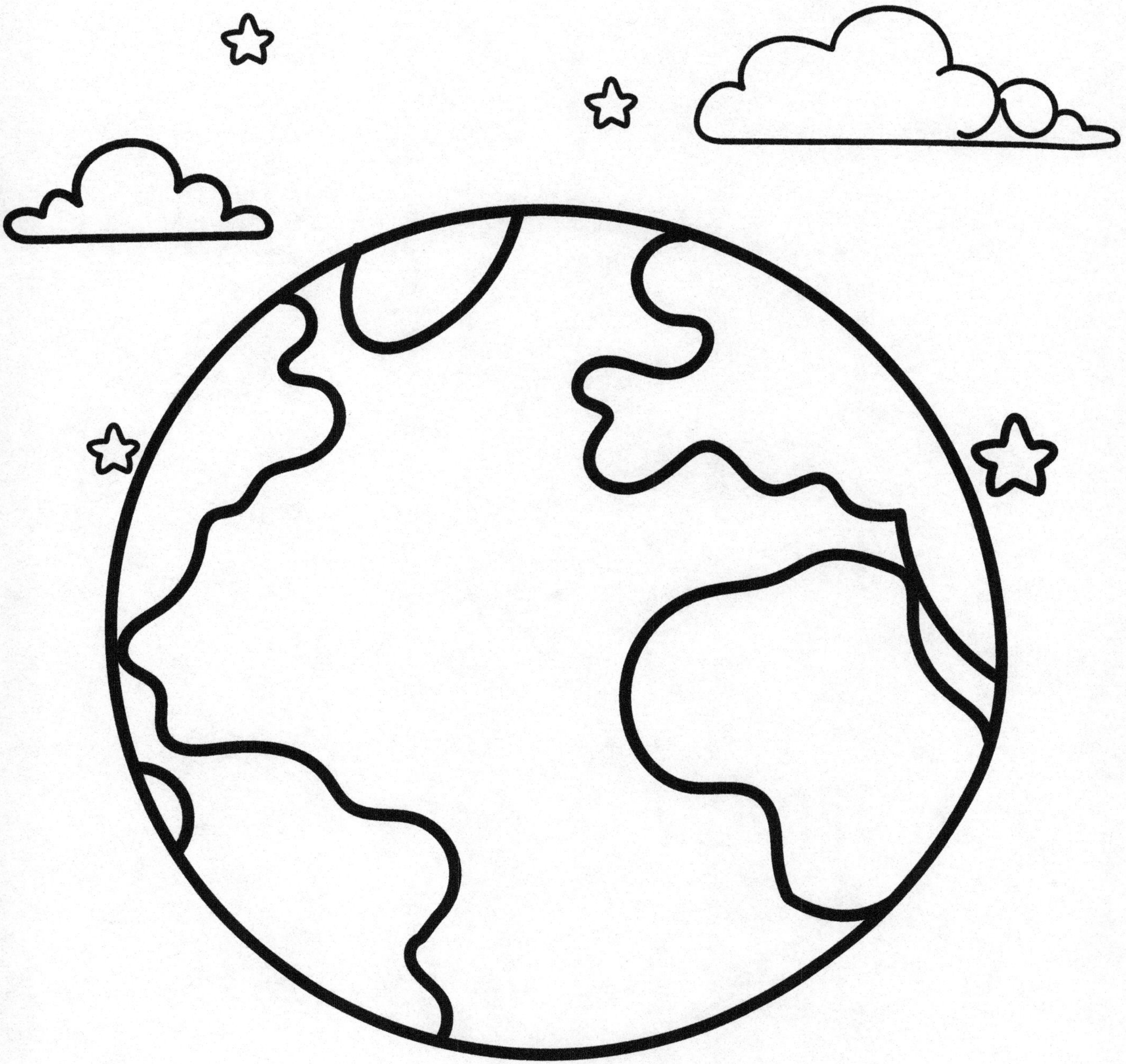

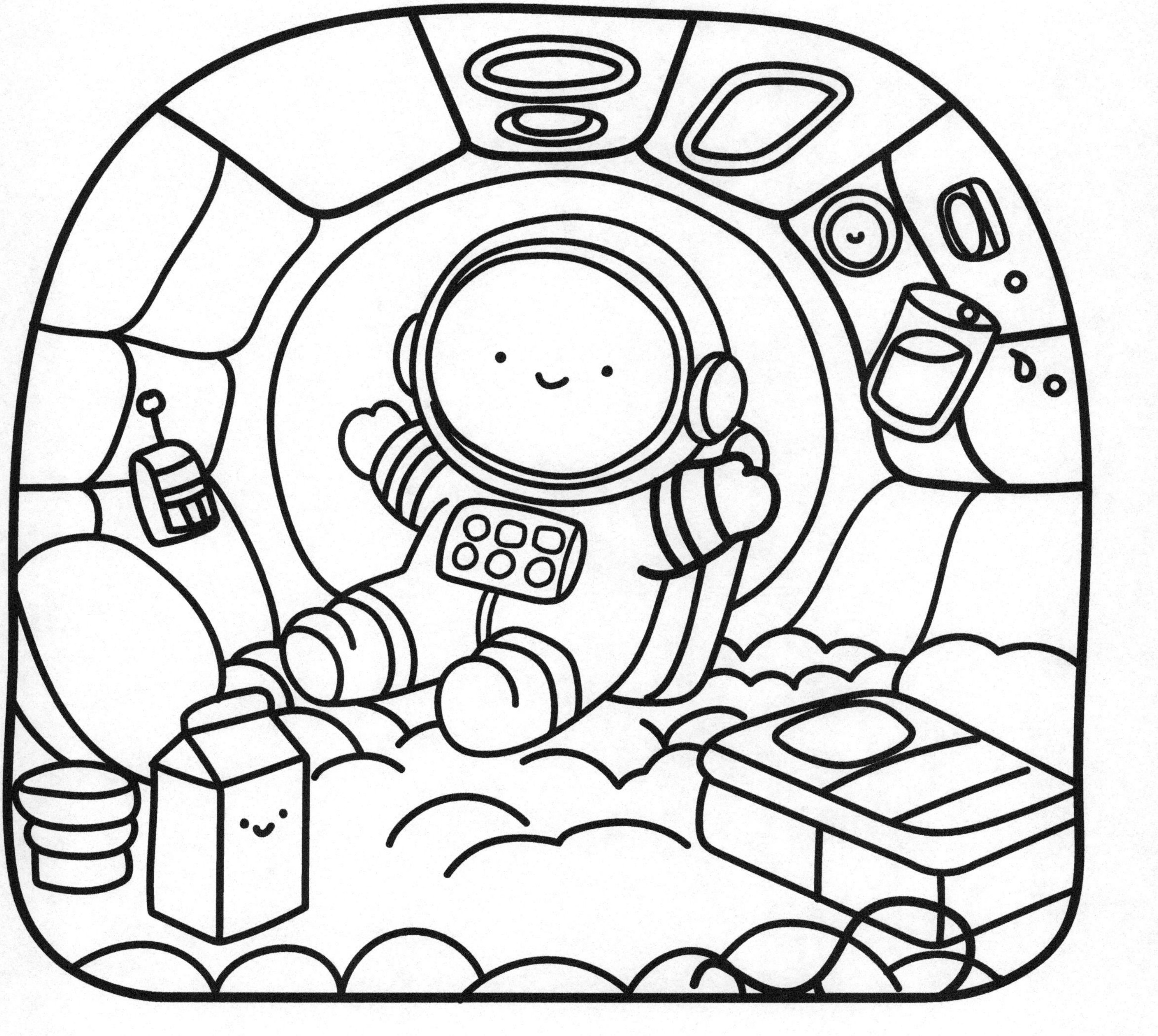

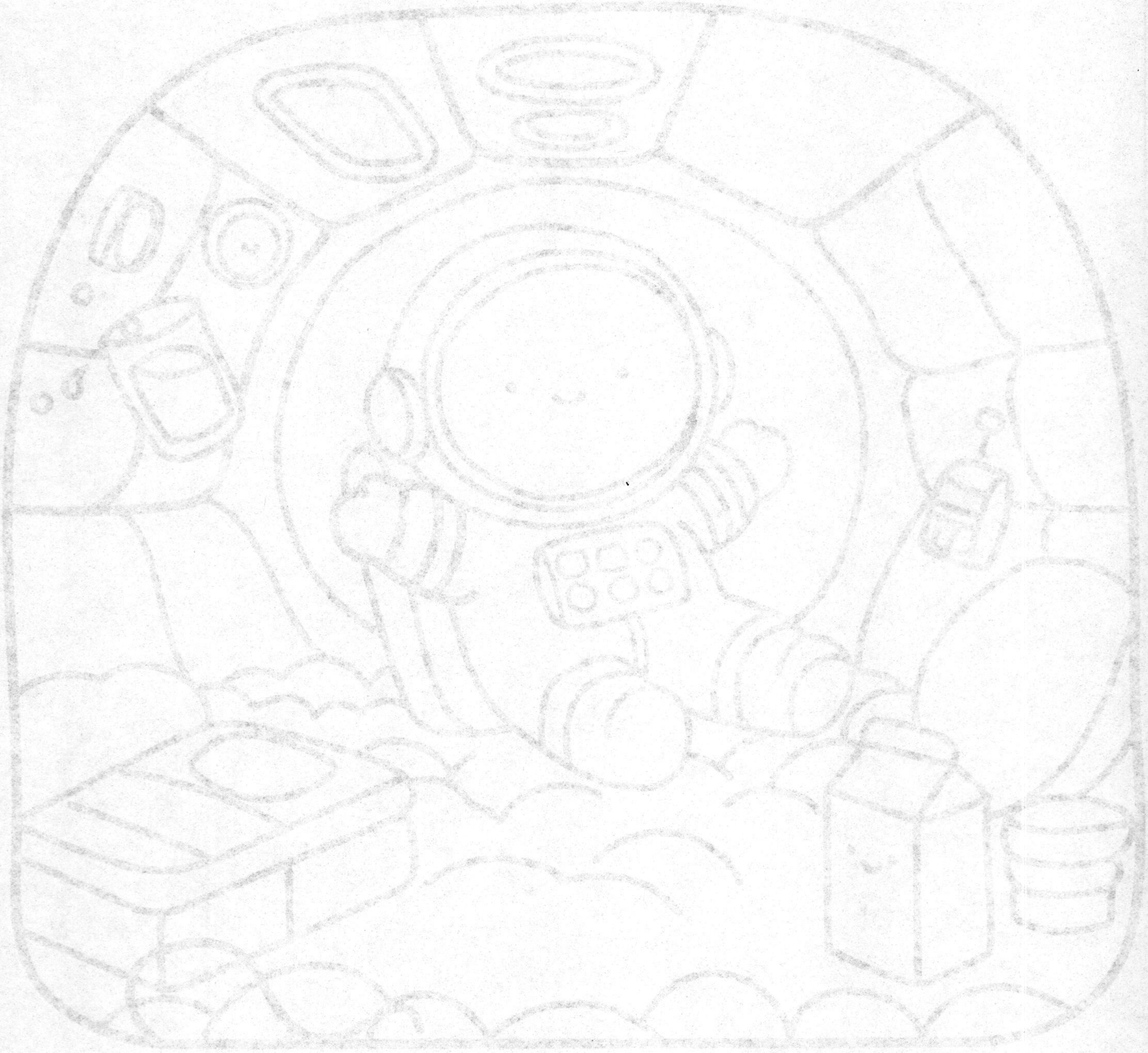

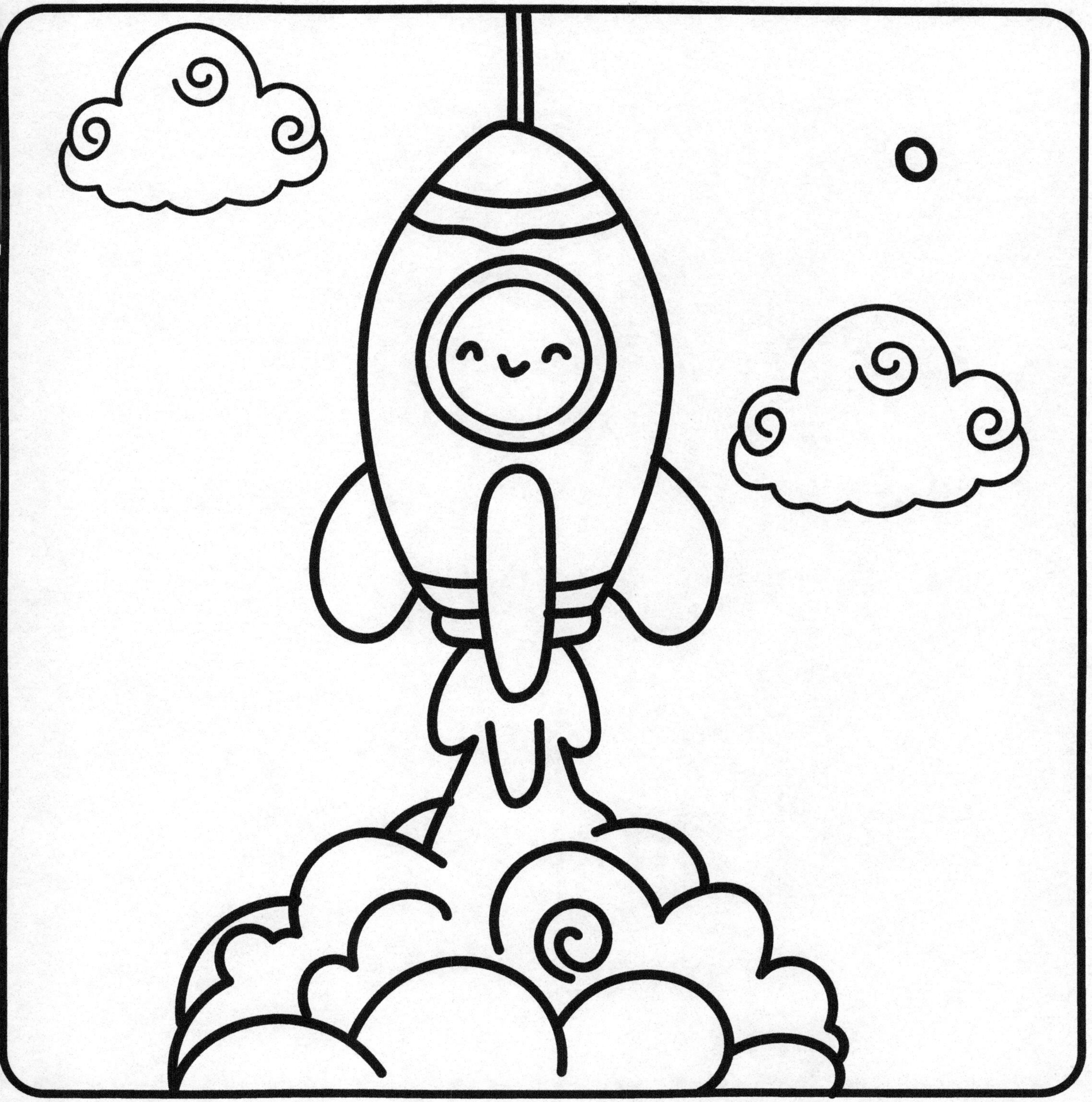

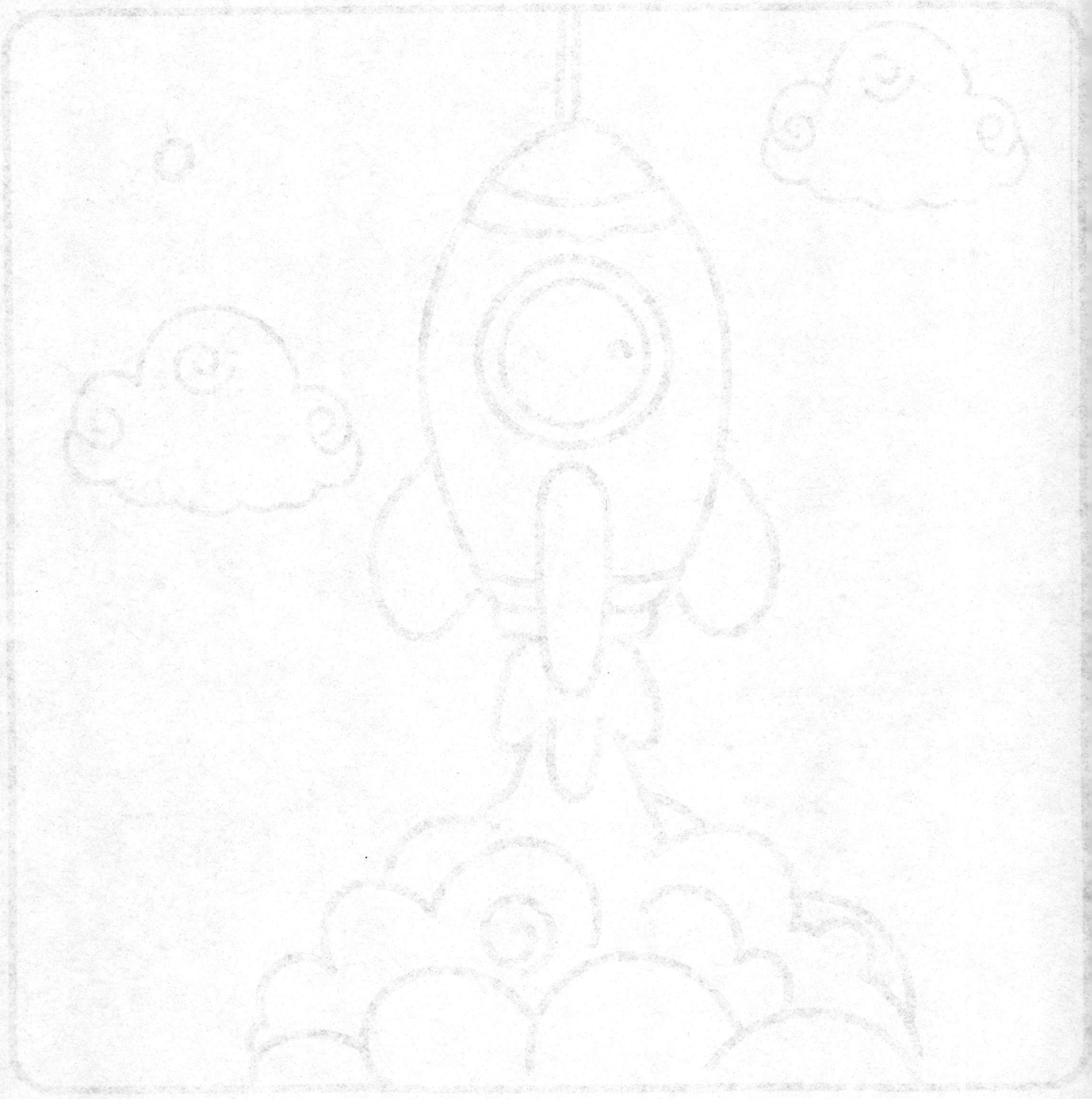

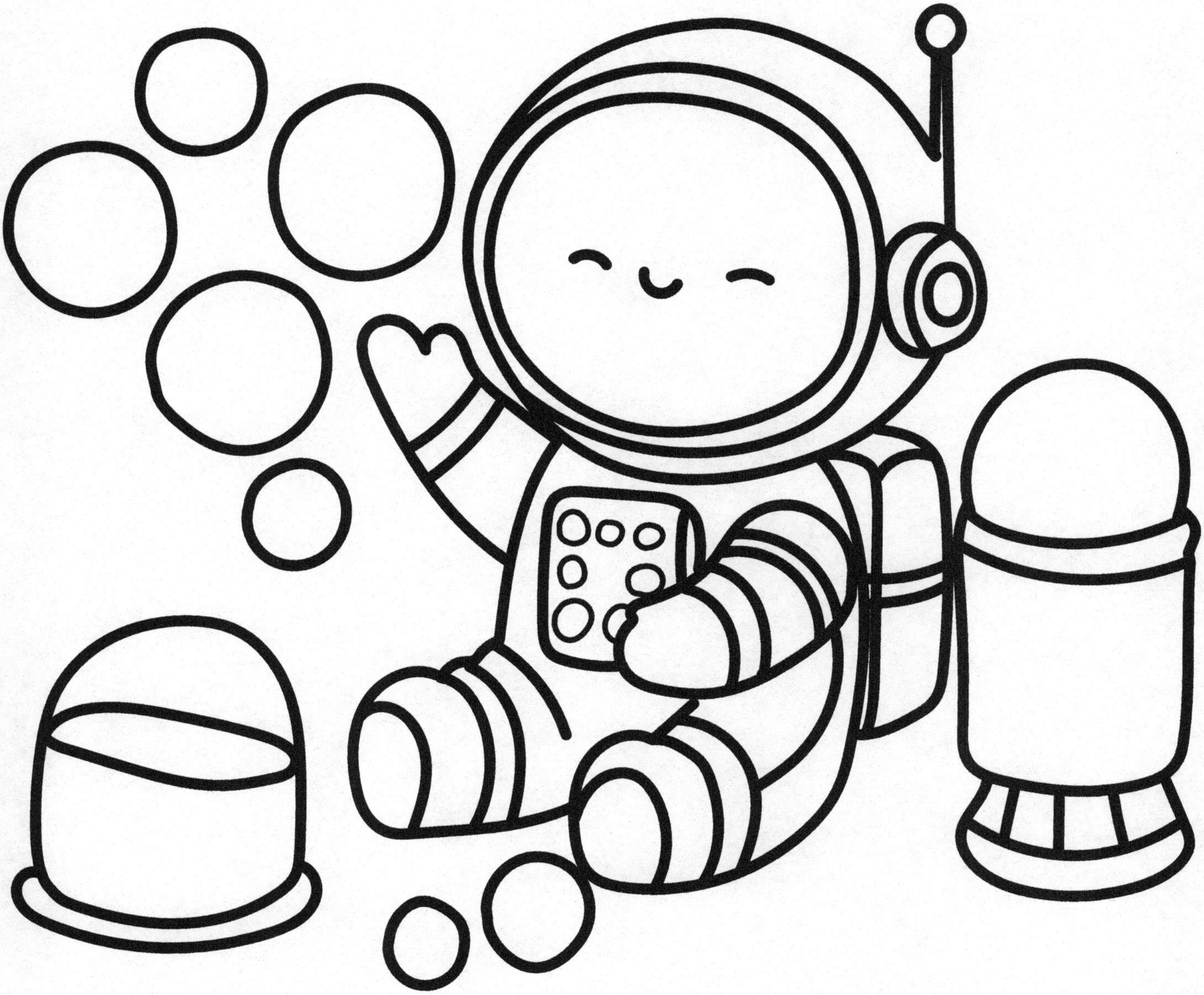

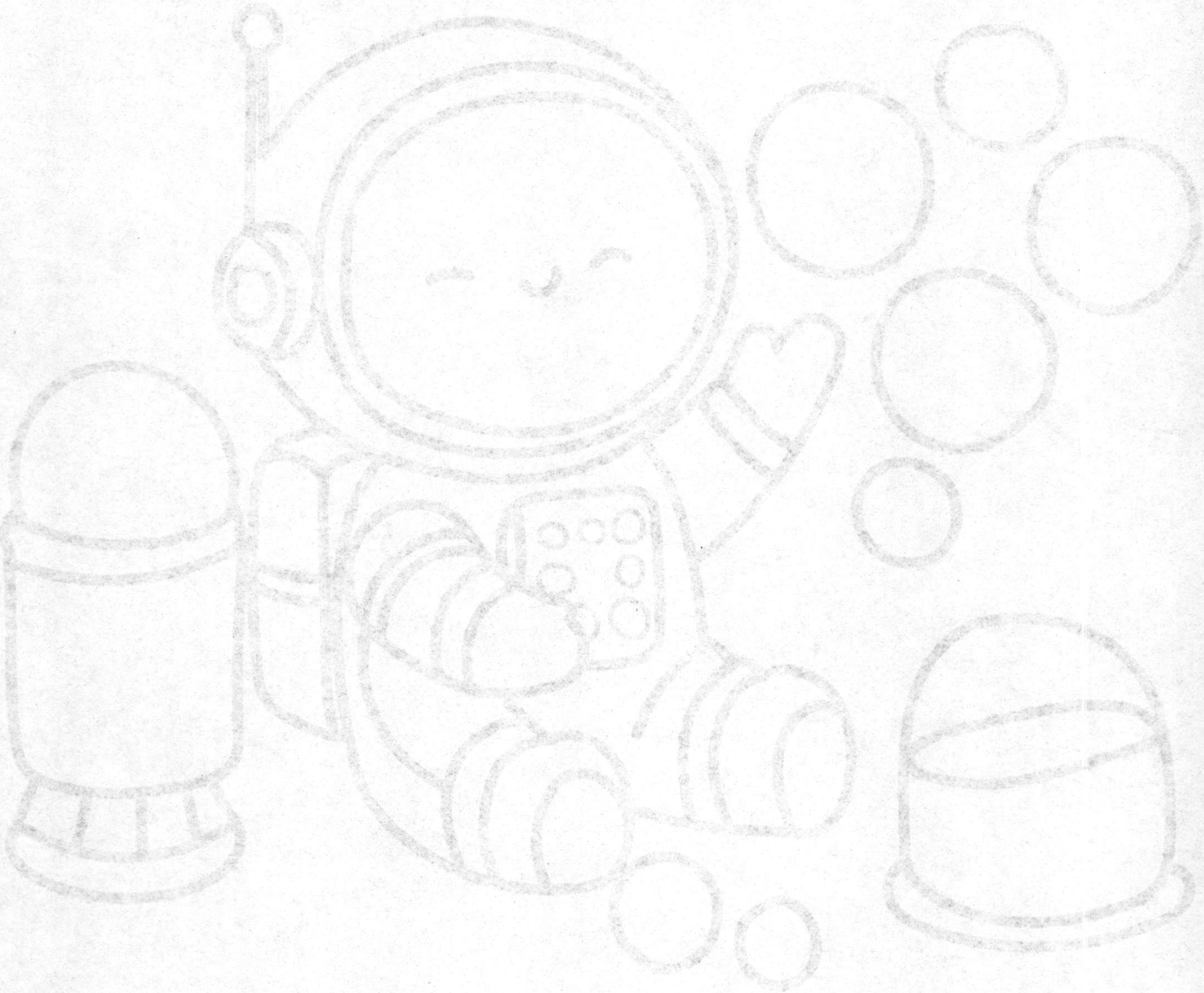

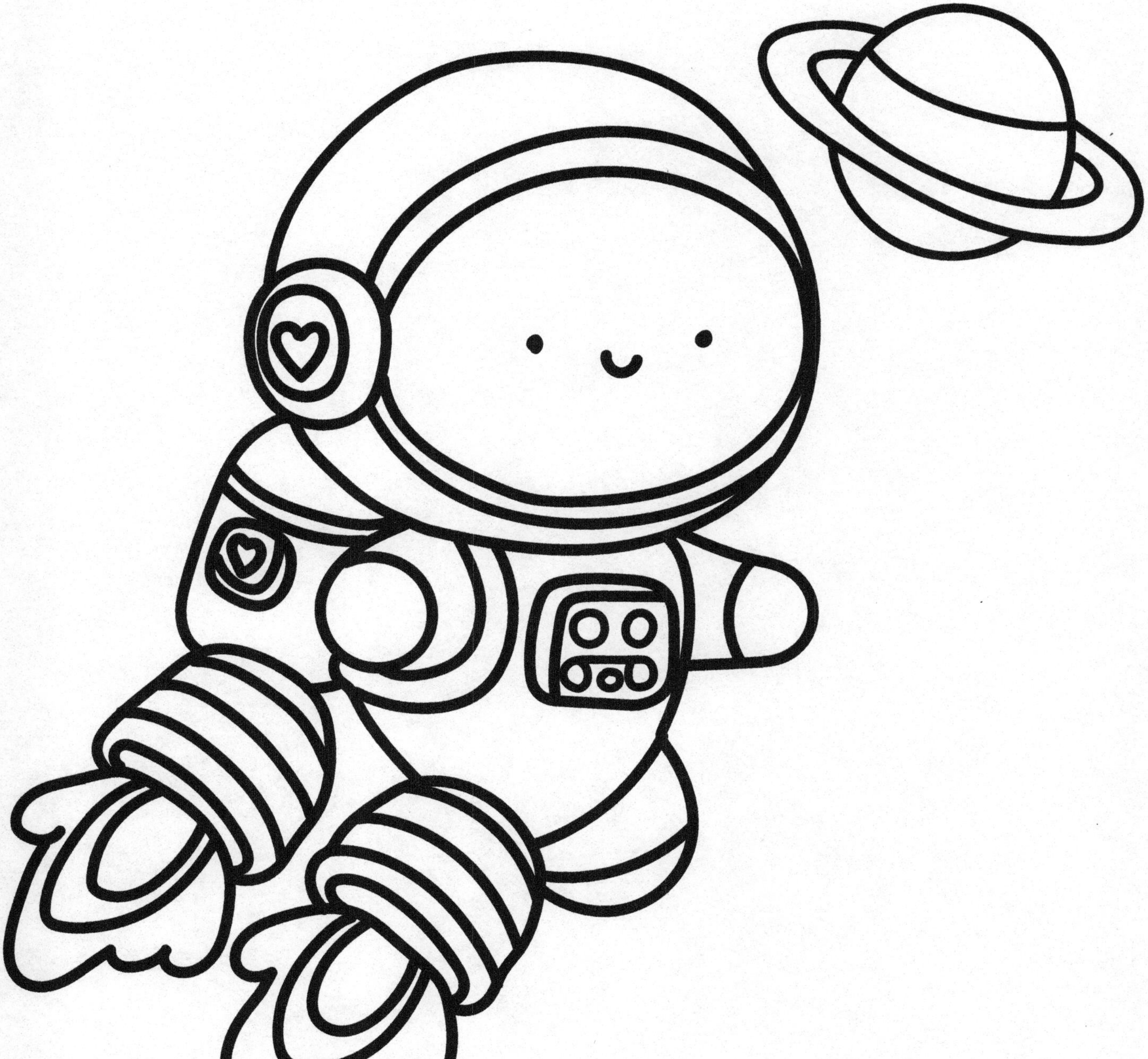

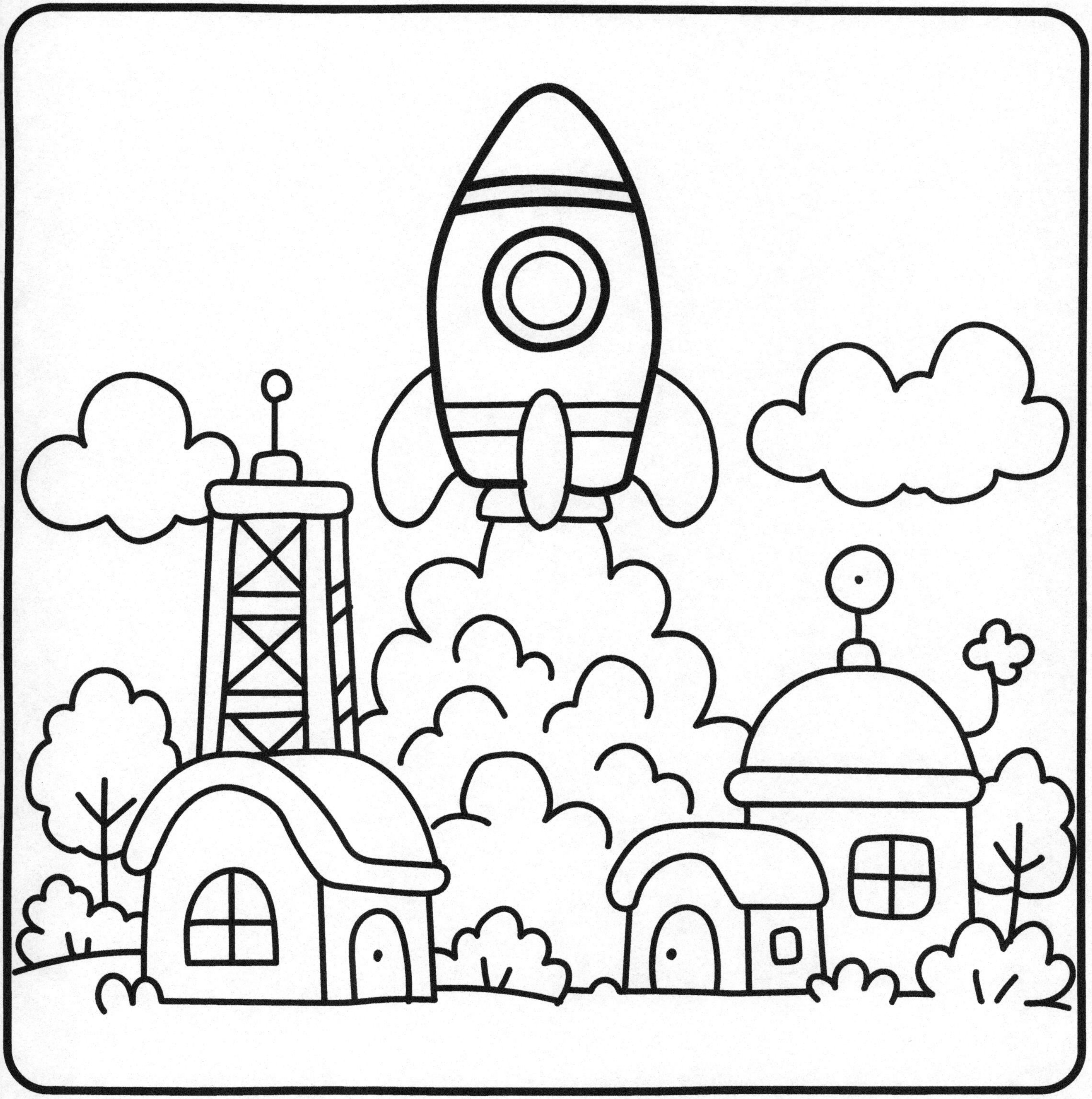

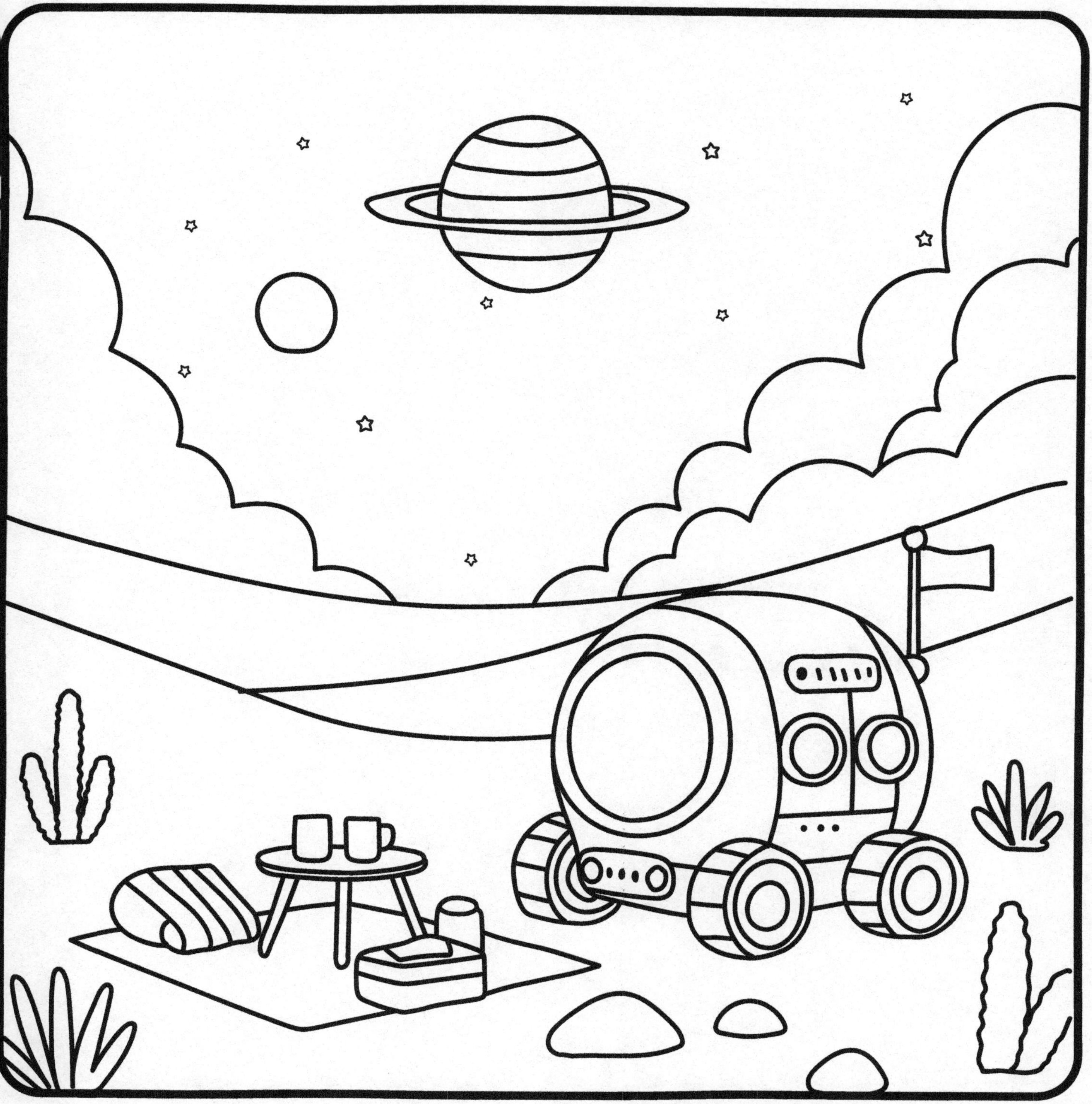

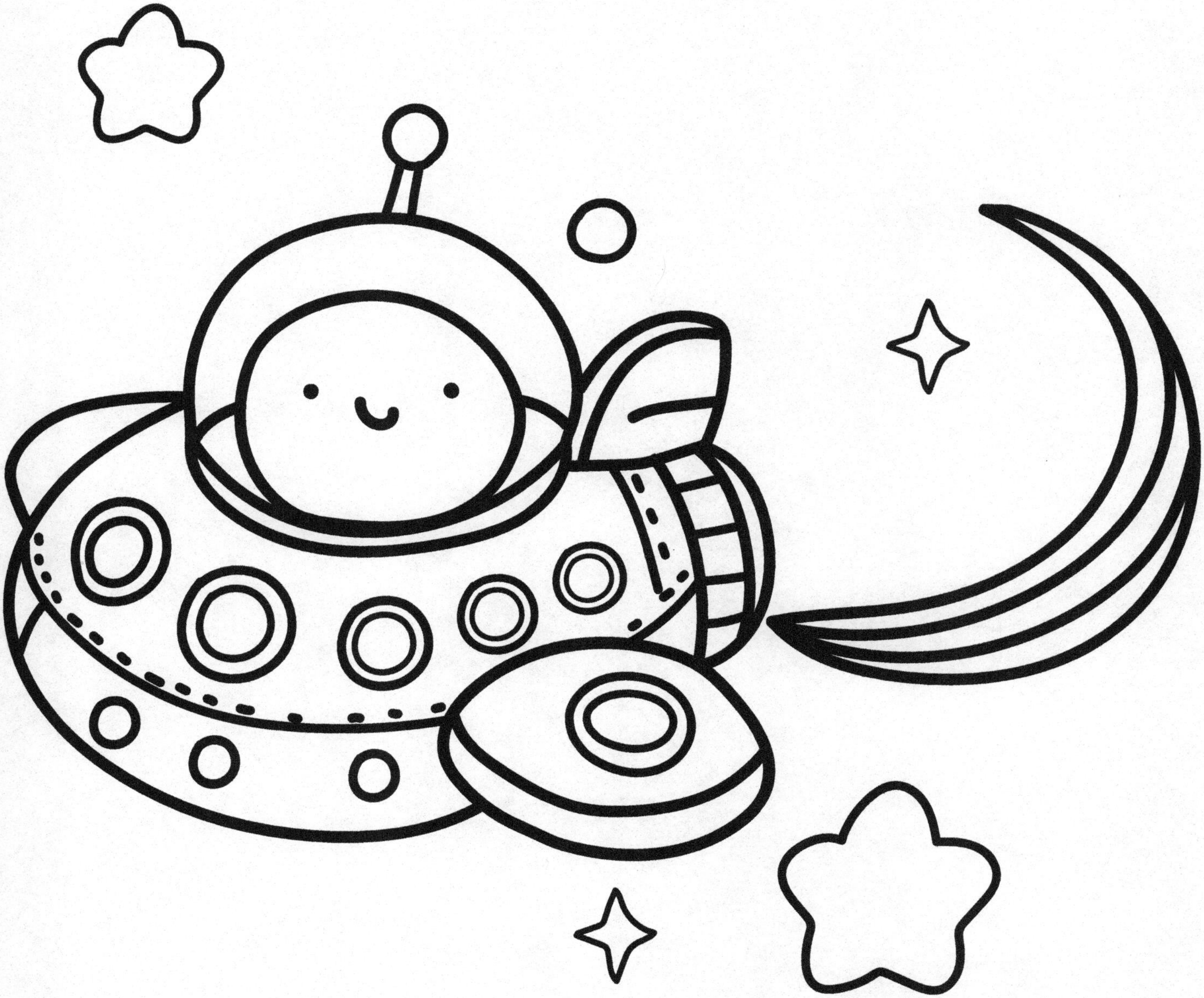

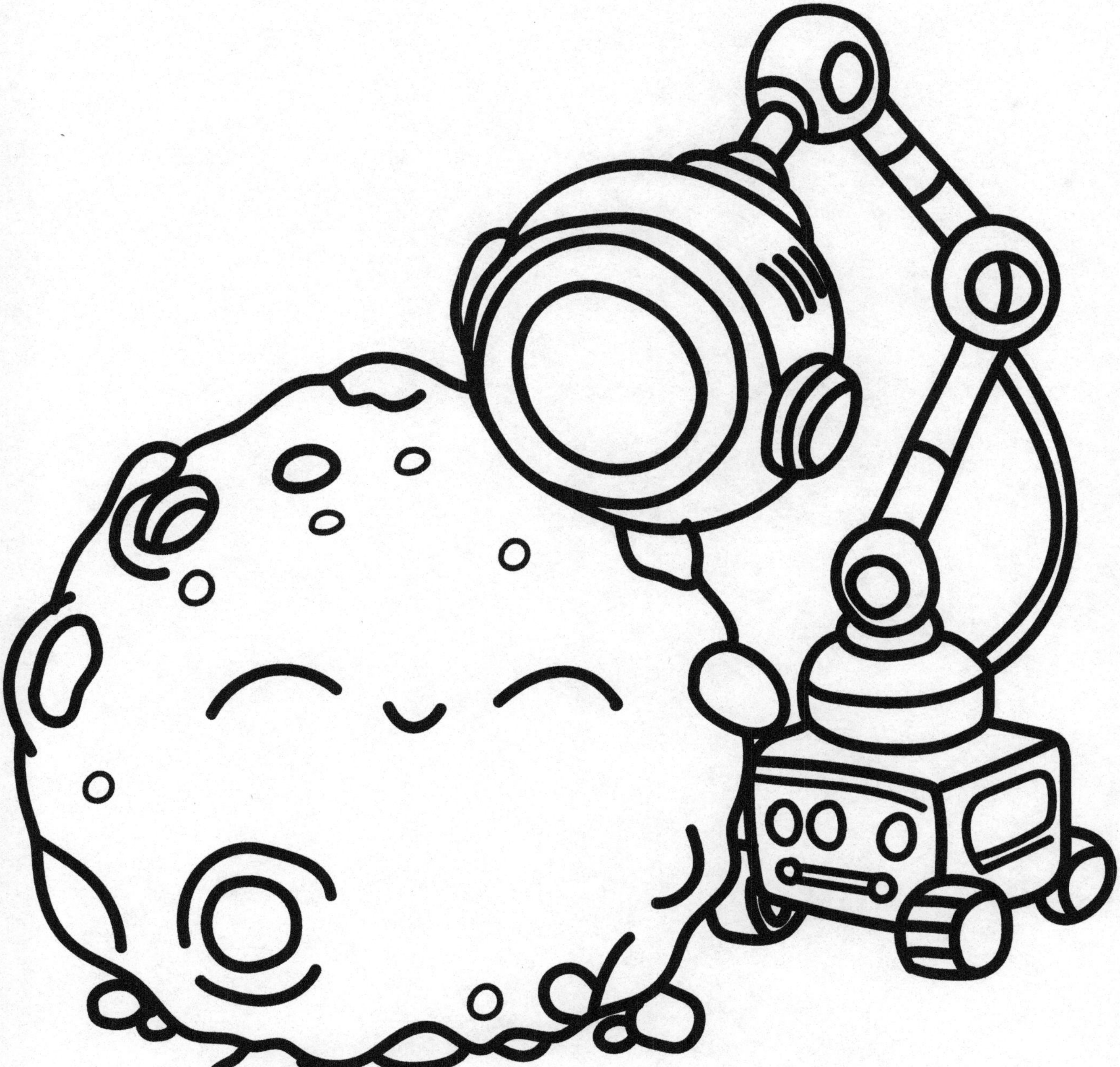

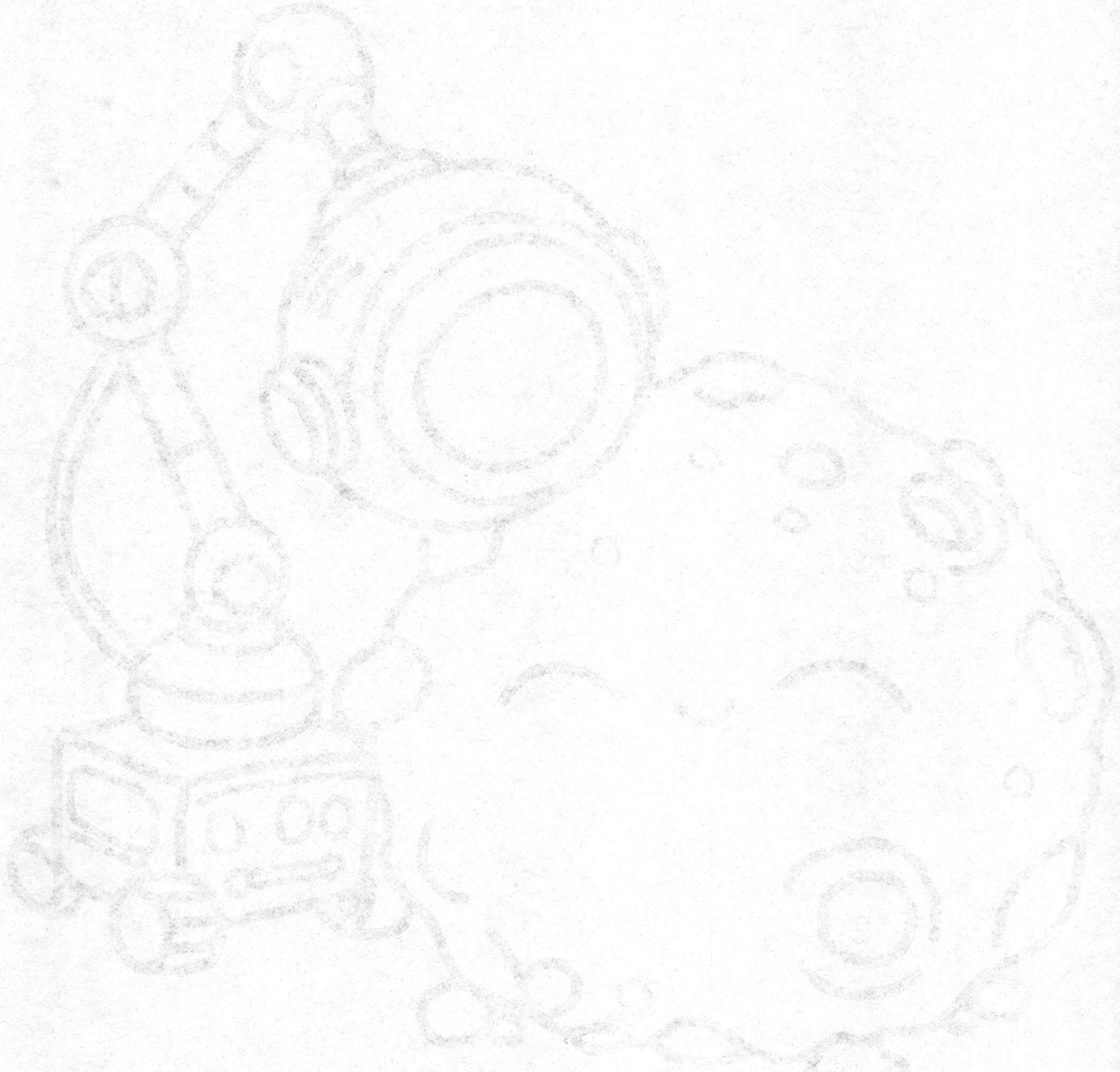

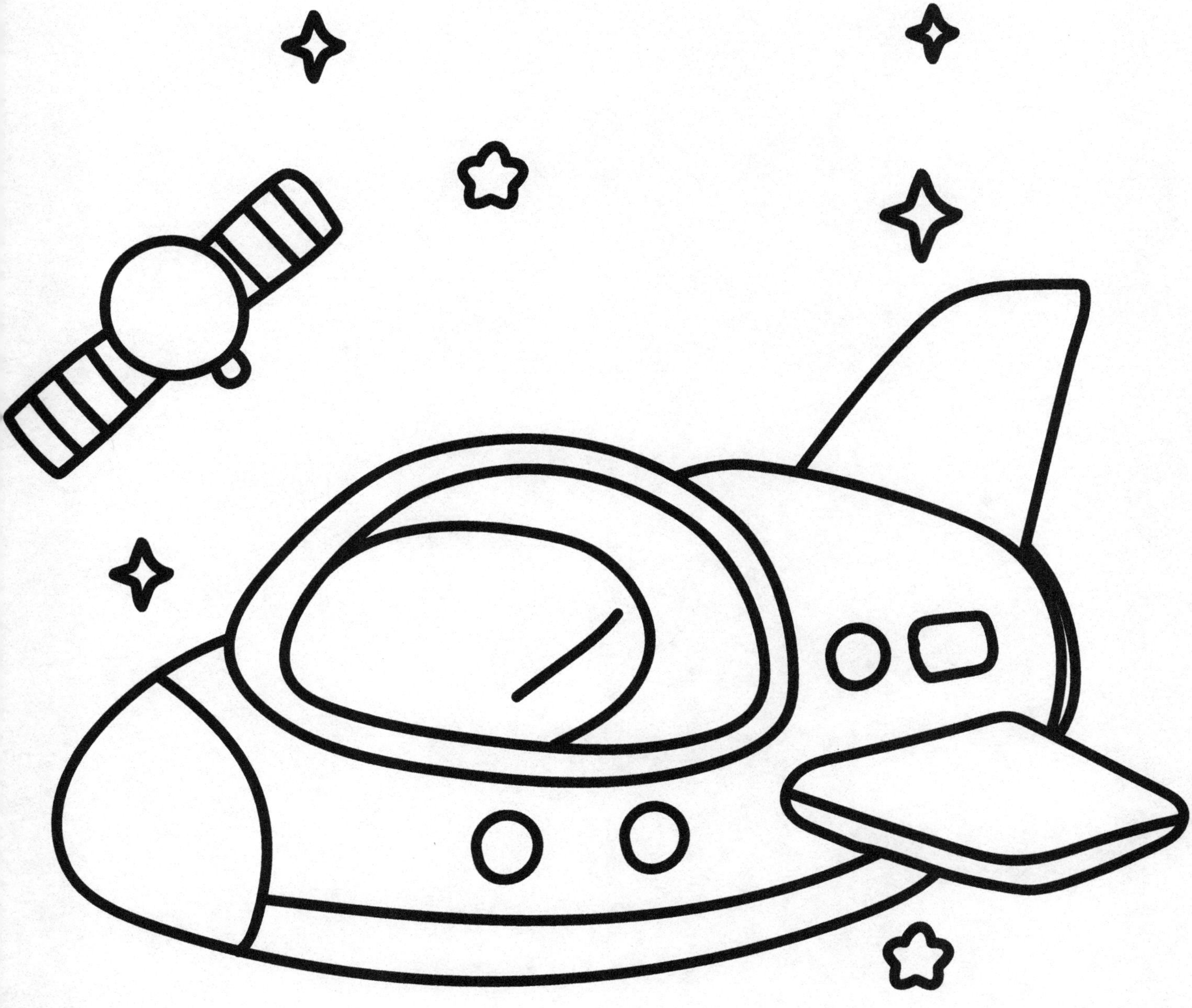

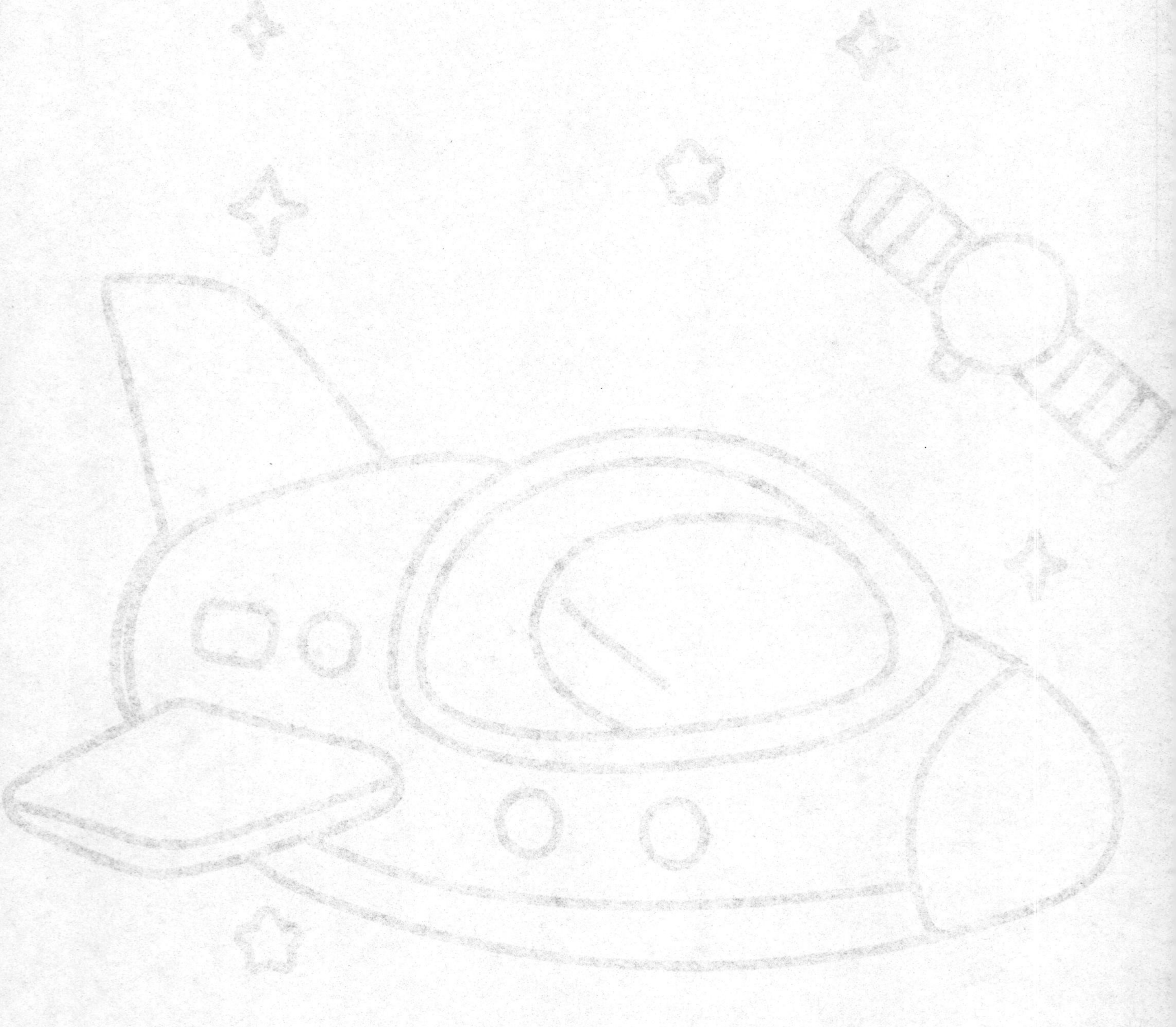

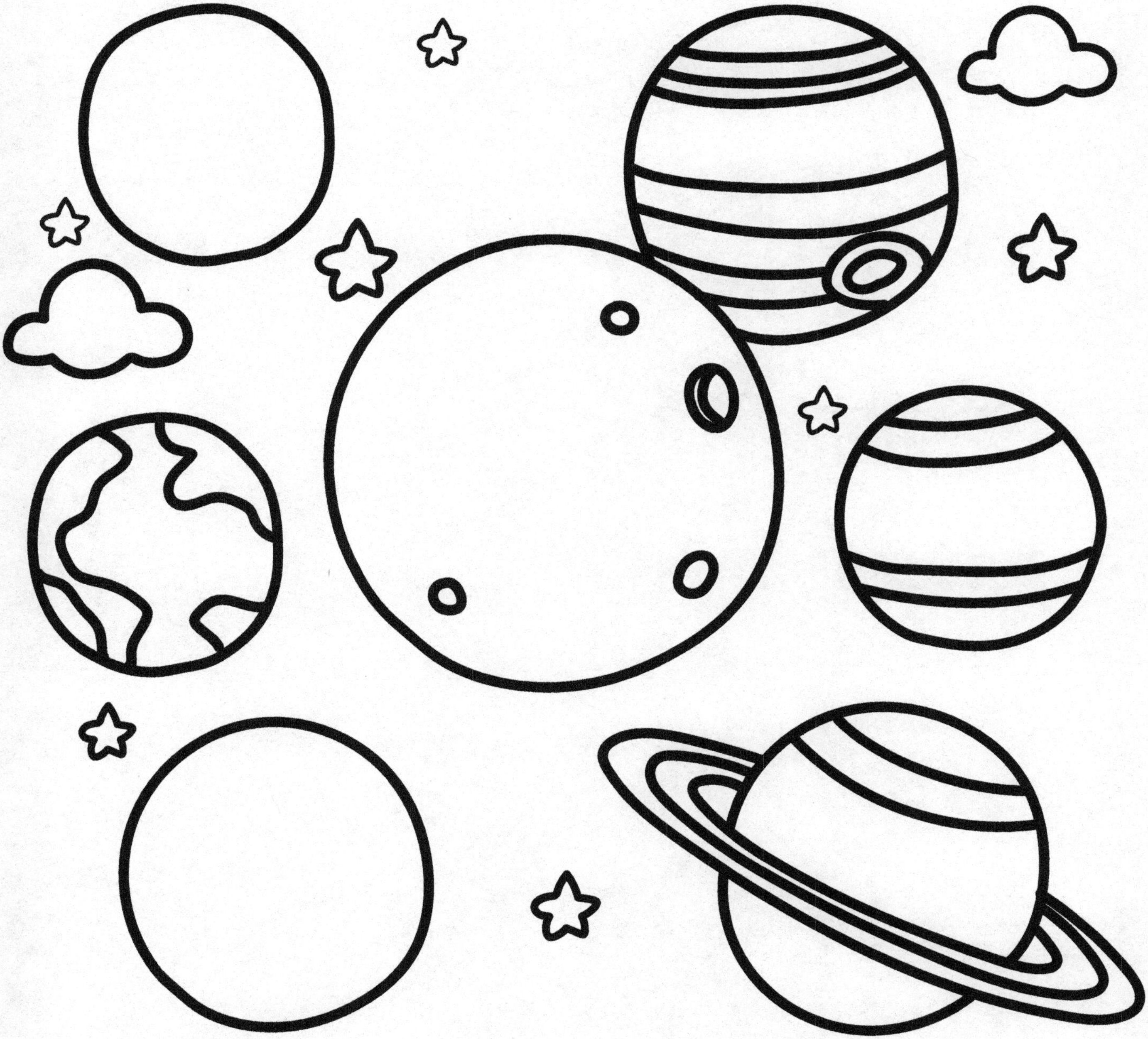

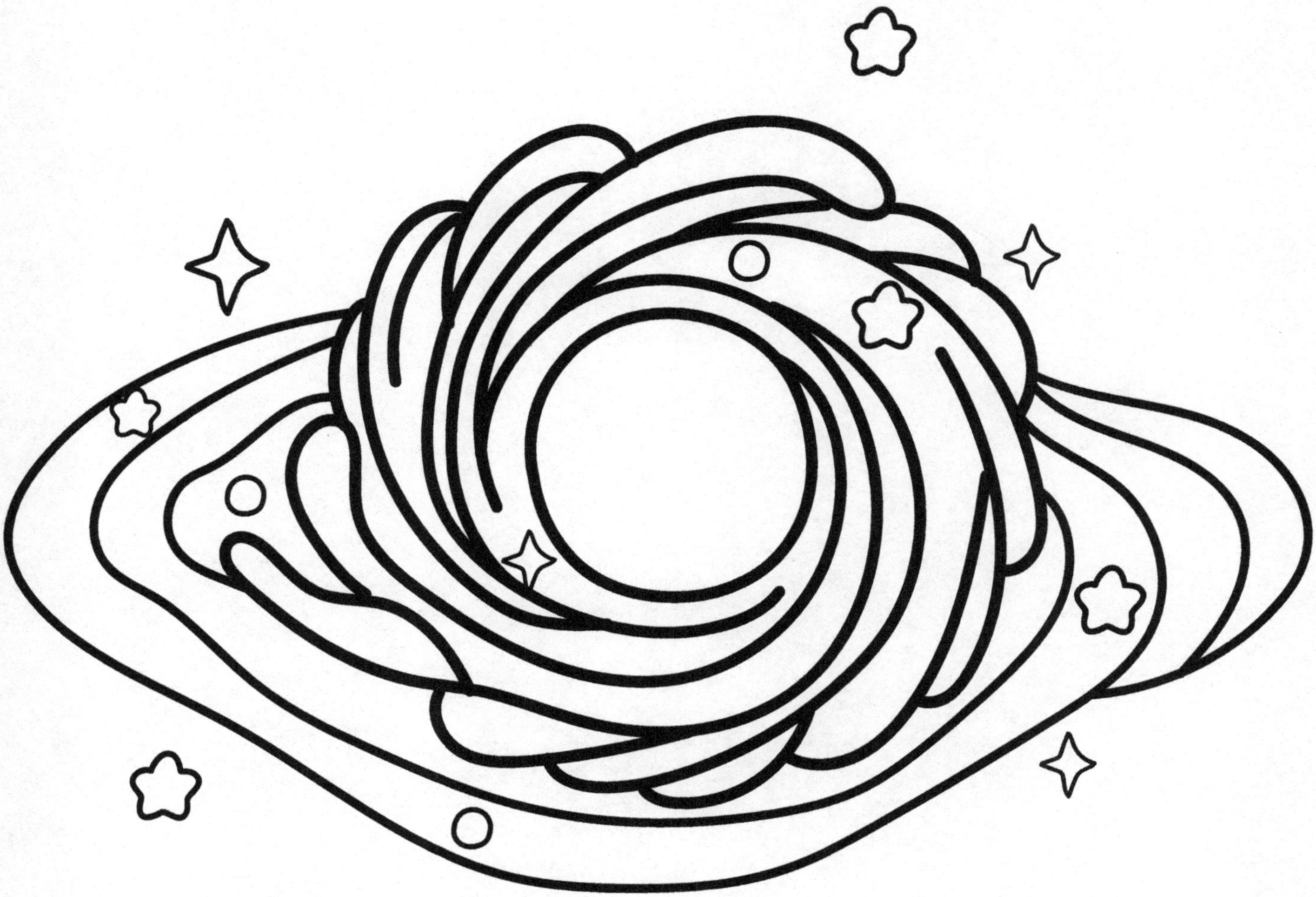

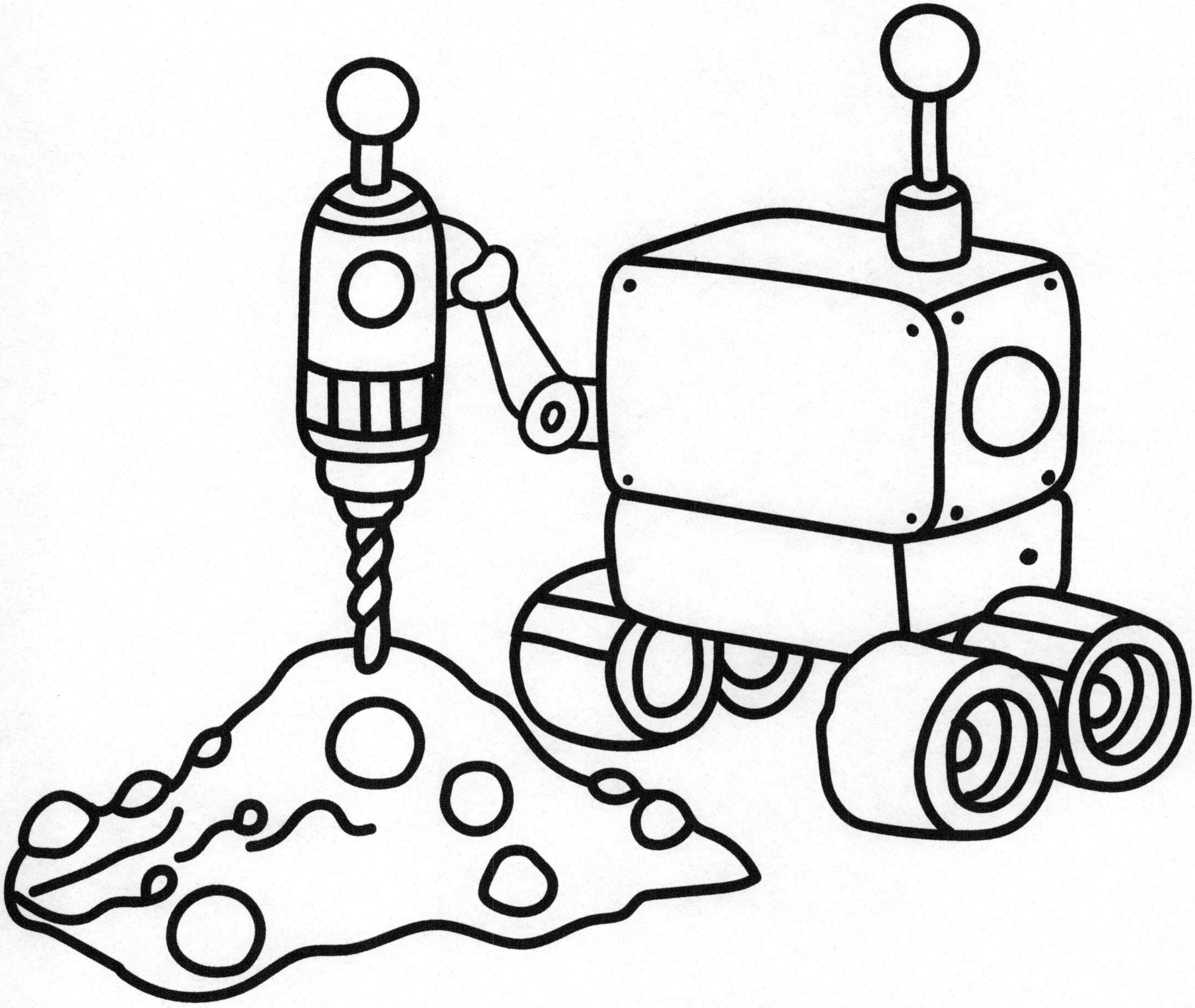

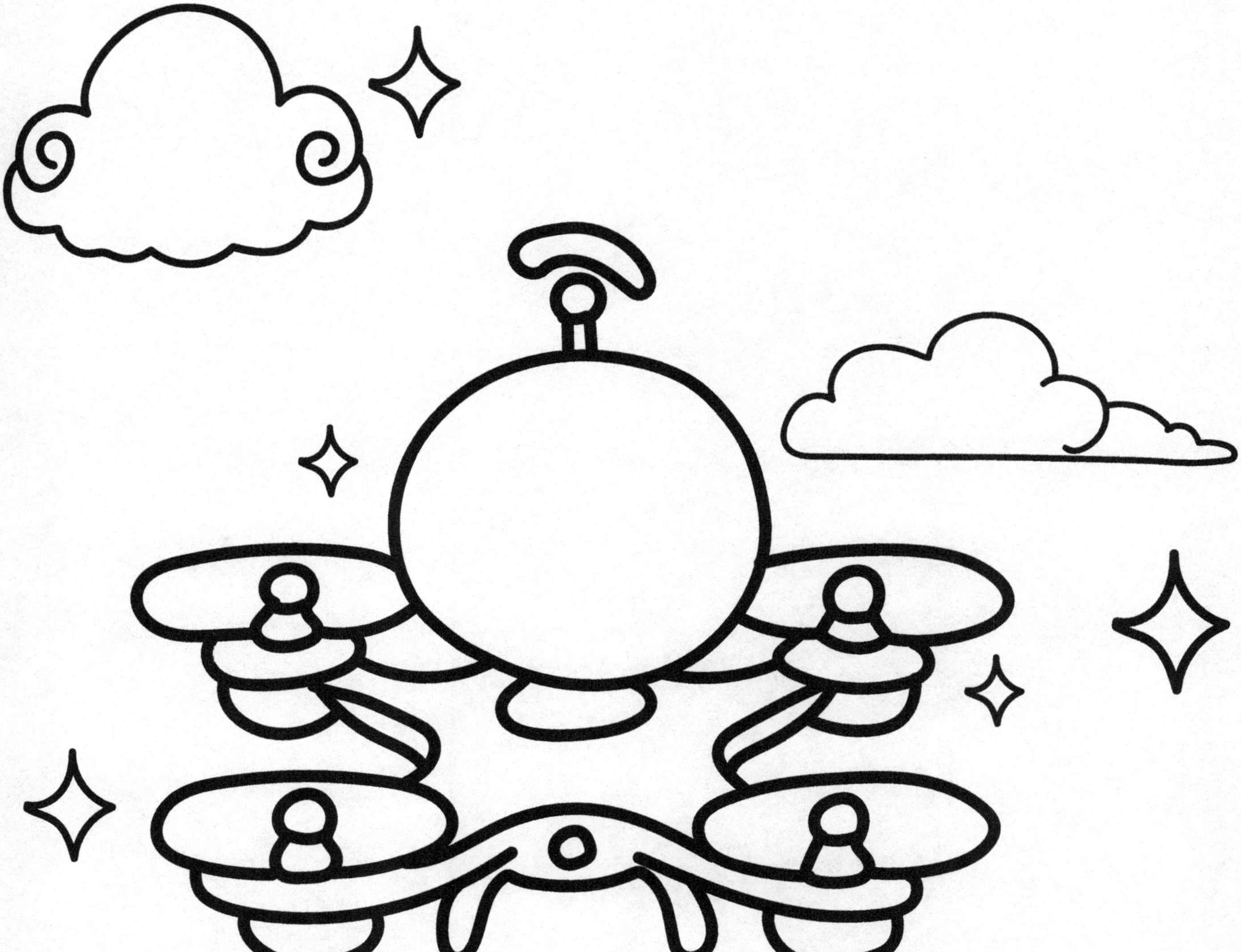

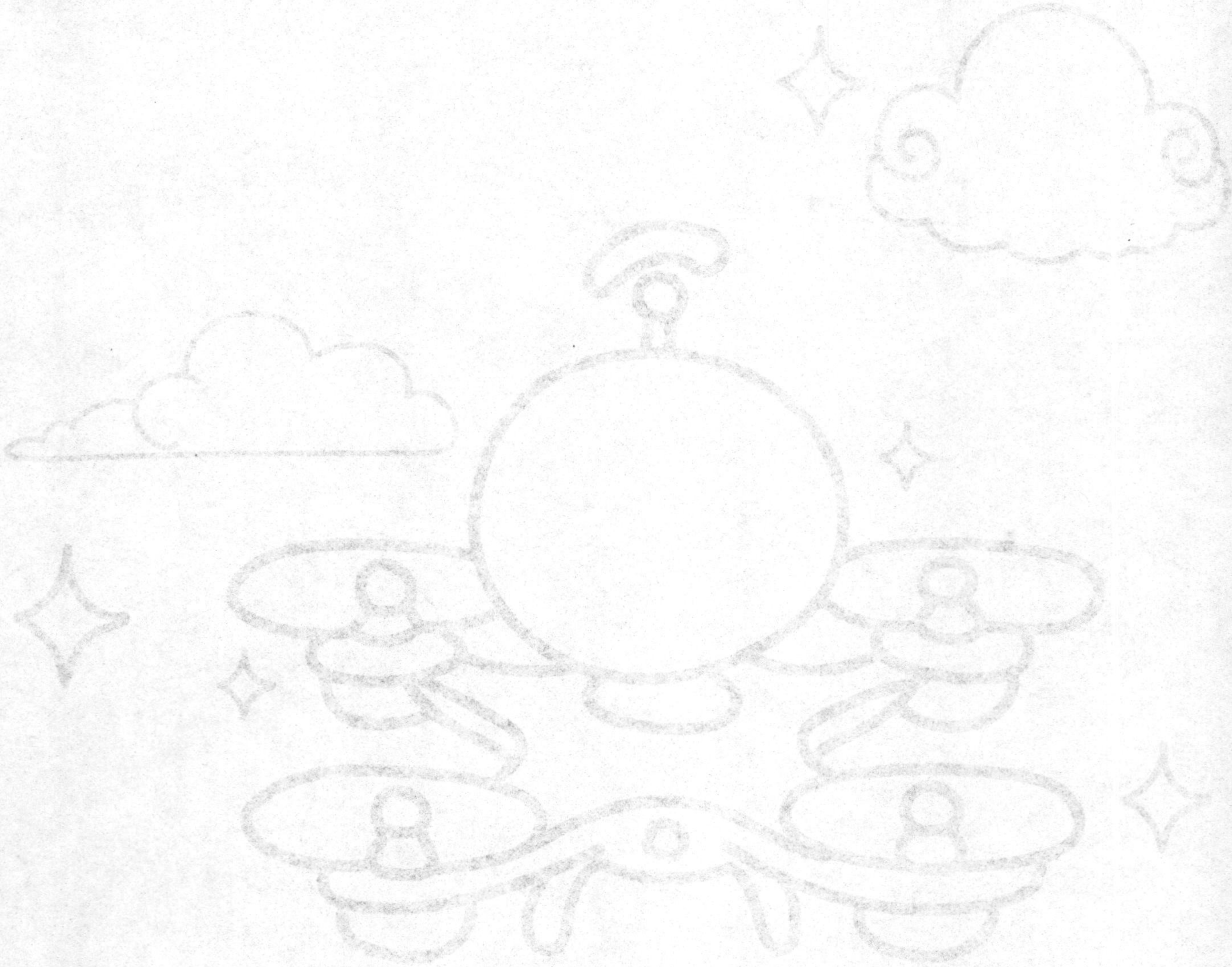

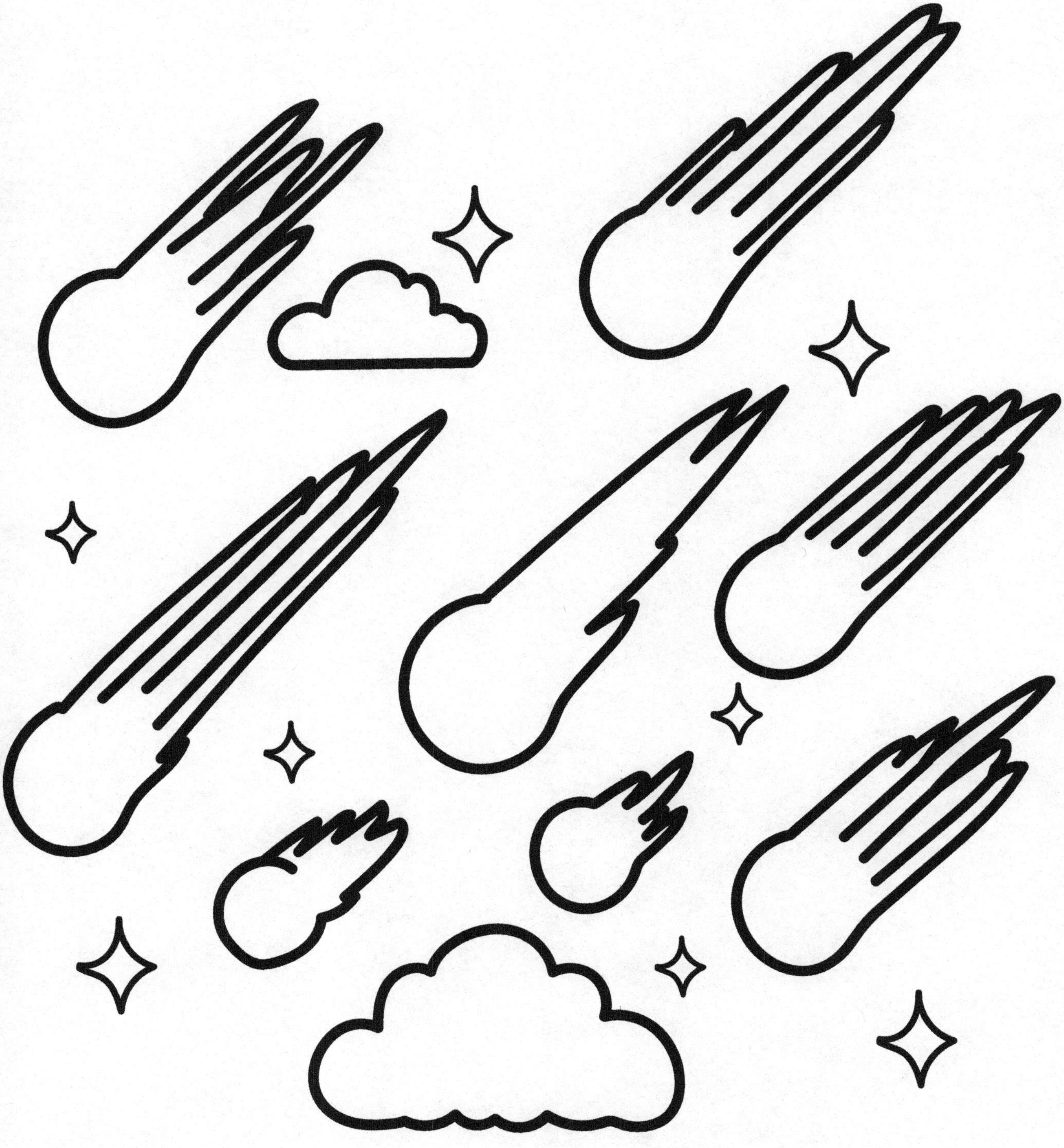

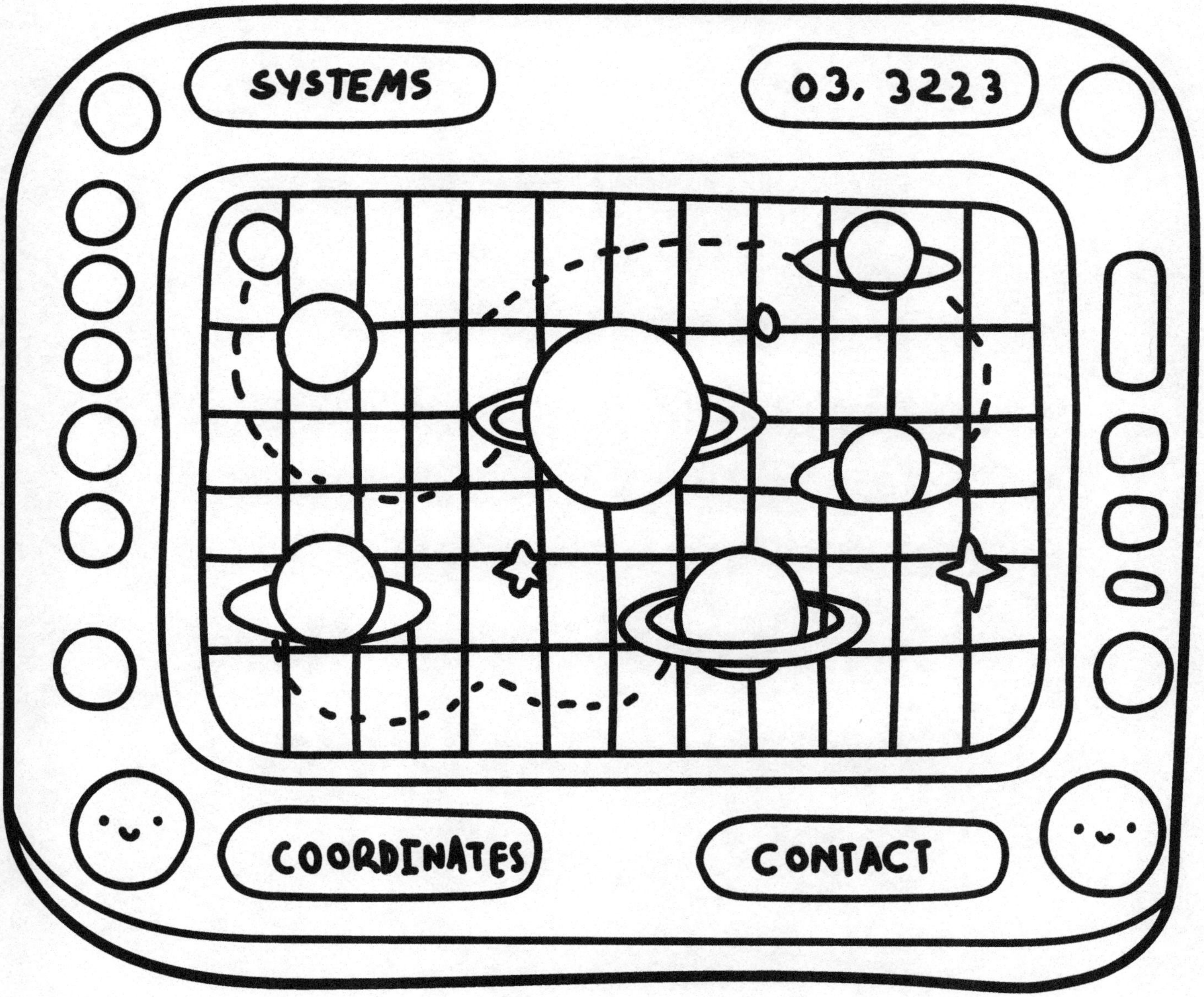
SYSTEMS
03, 3223
COORDINATES
CONTACT

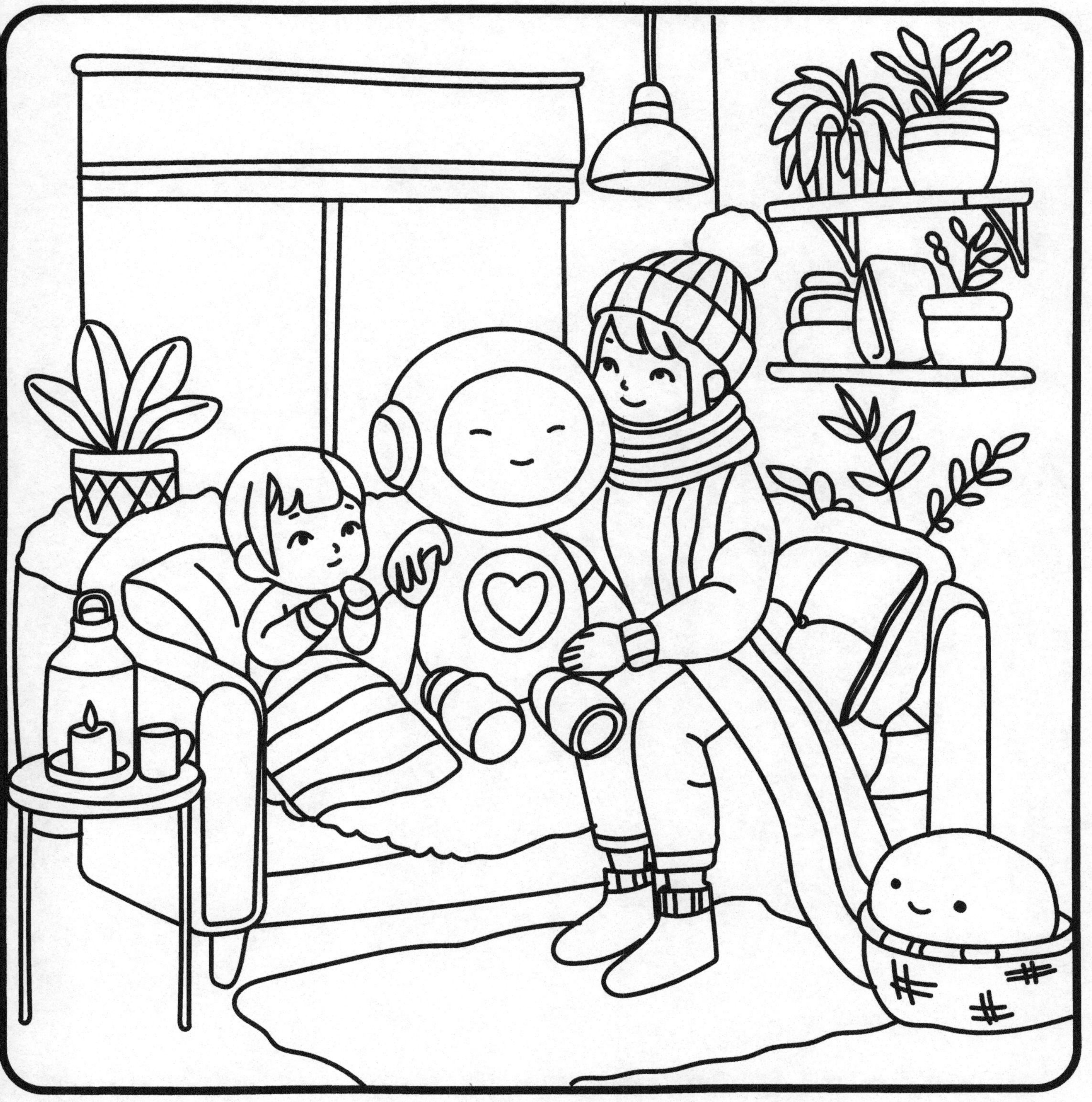

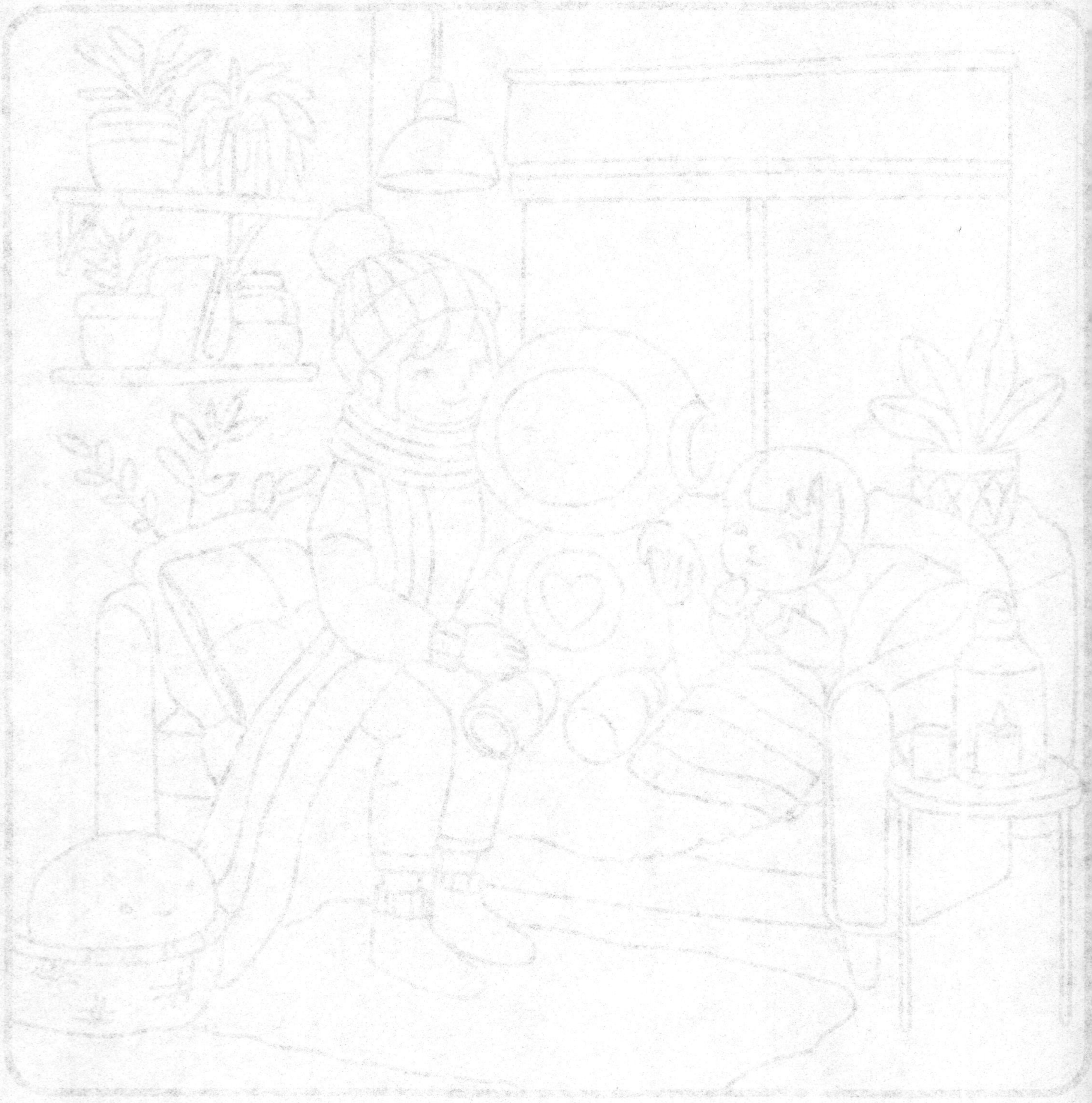

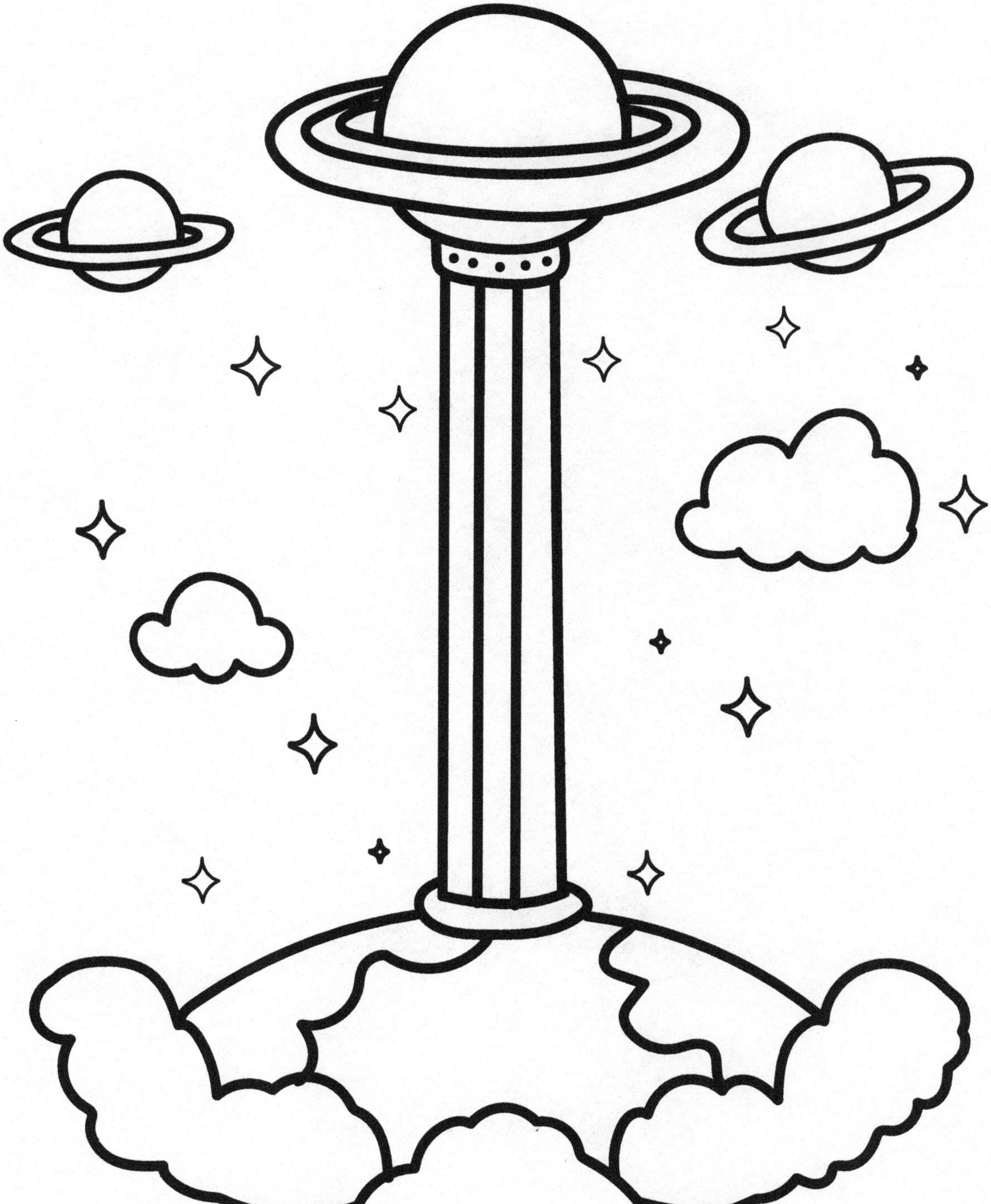

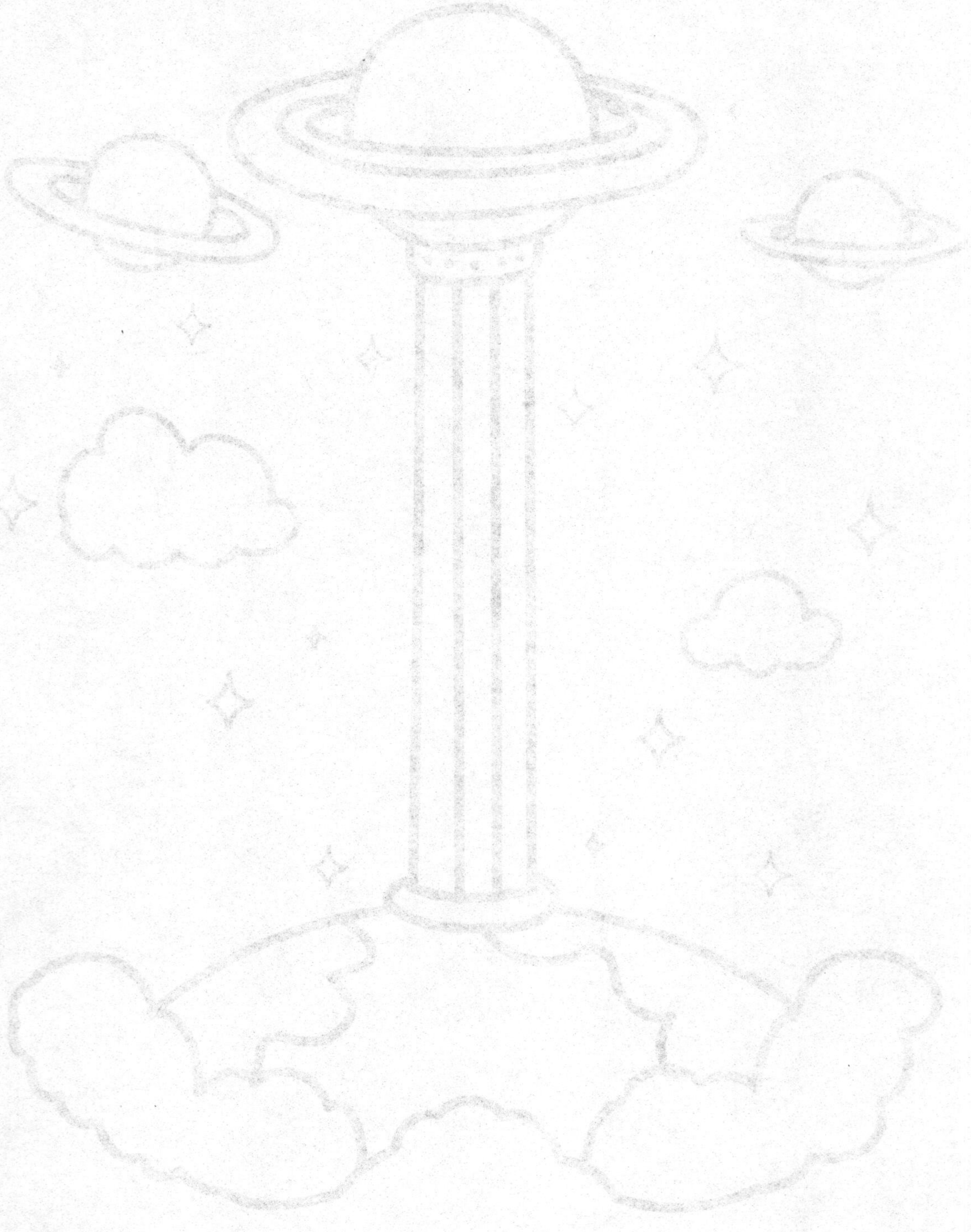

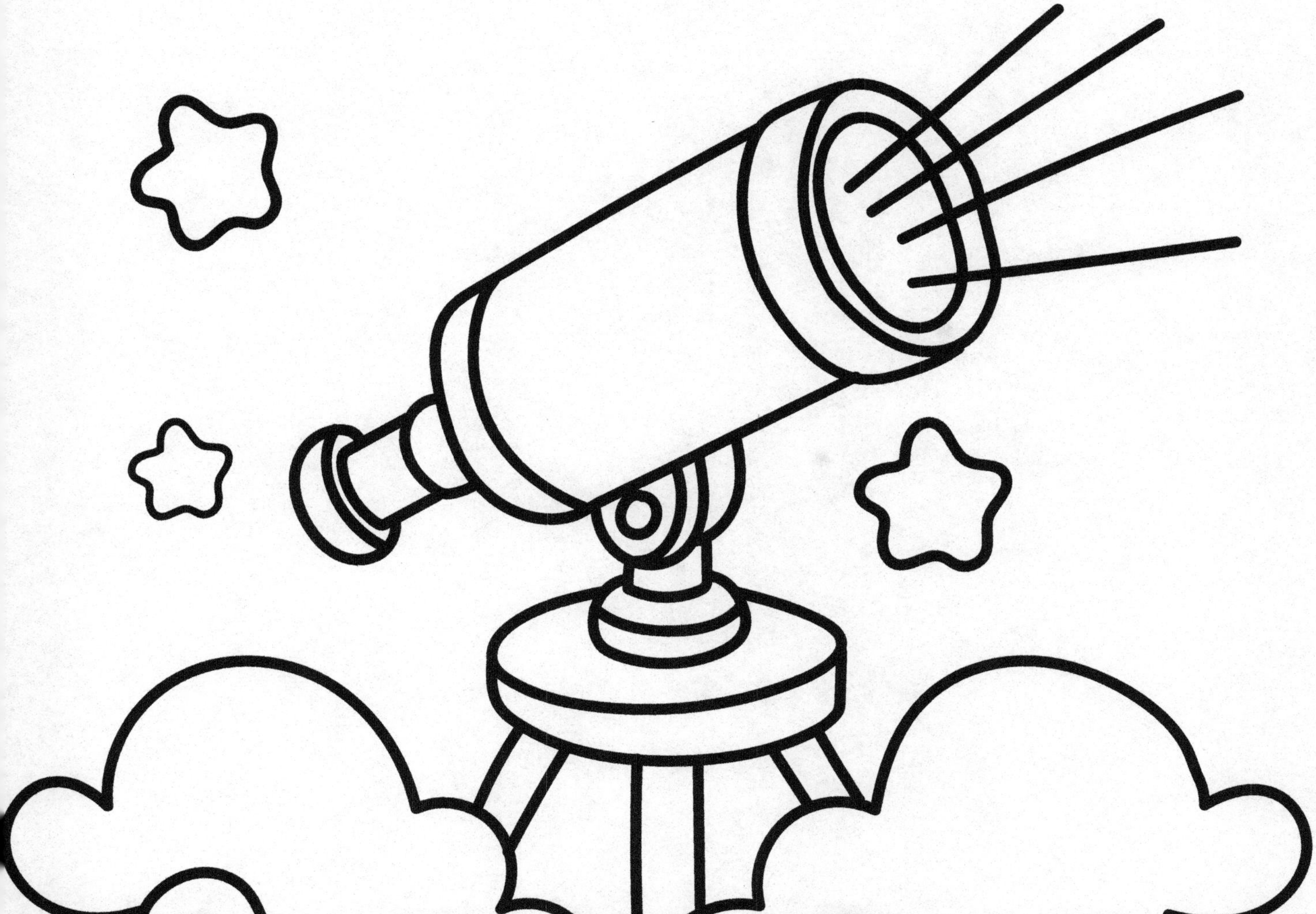

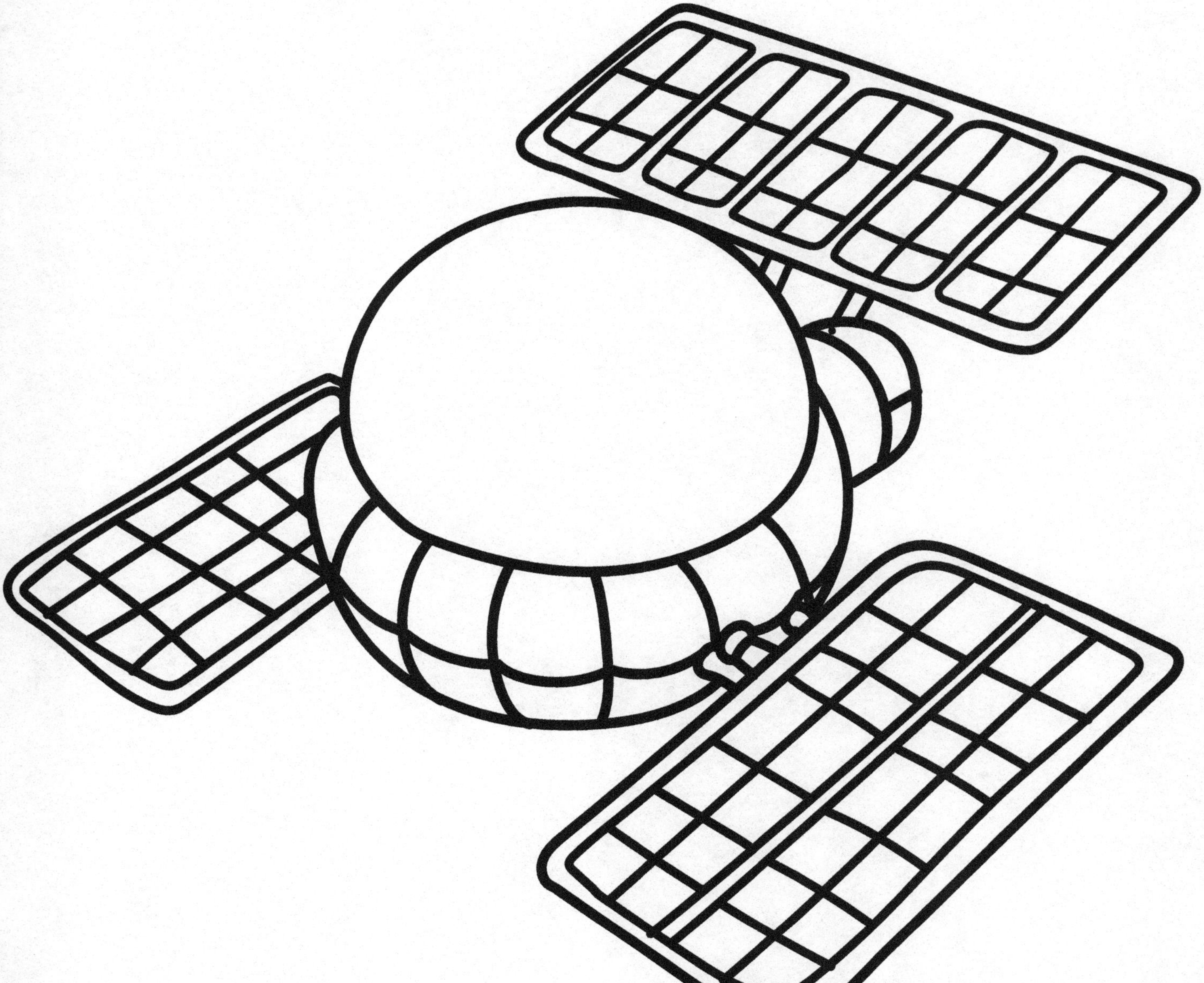

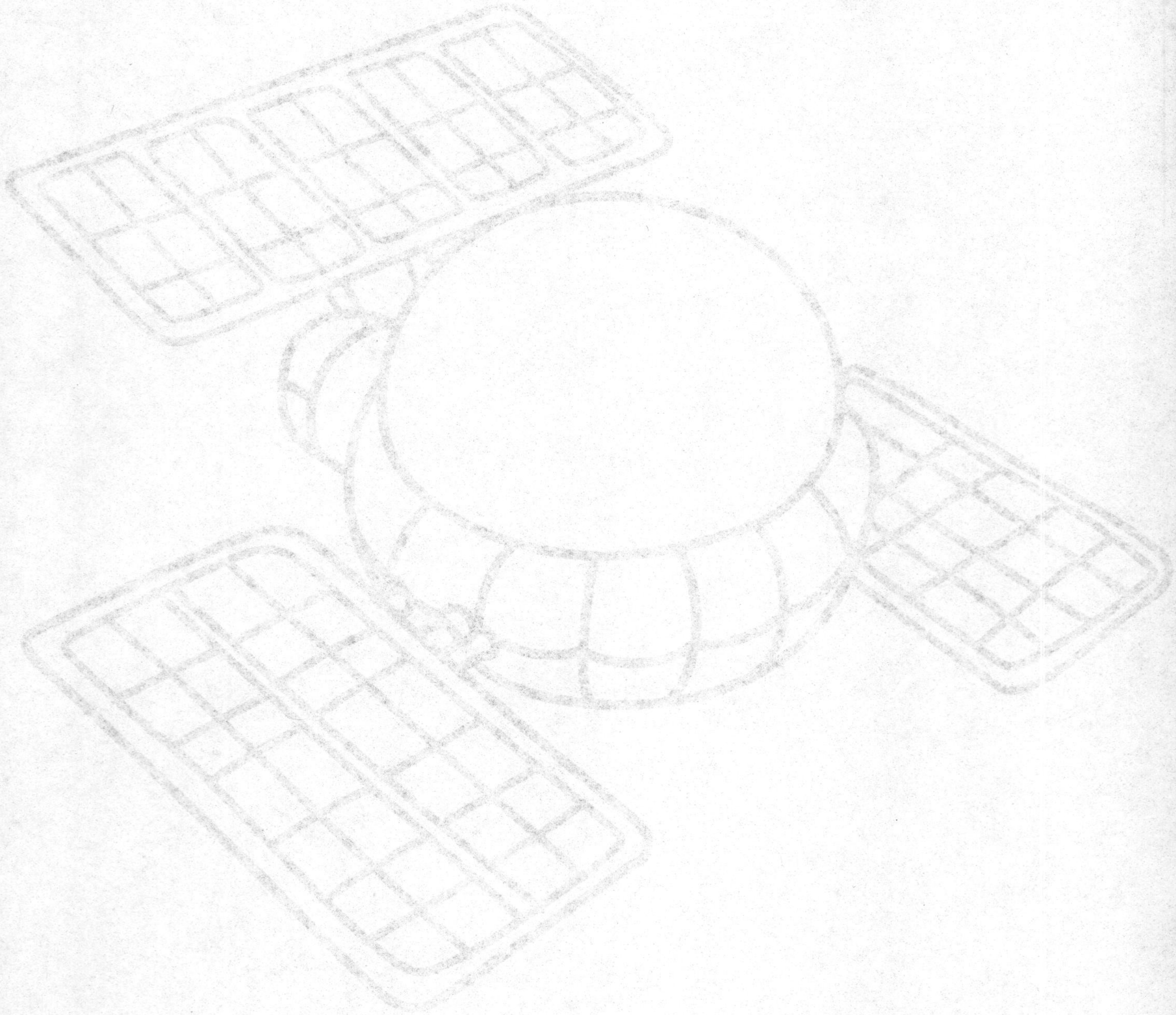

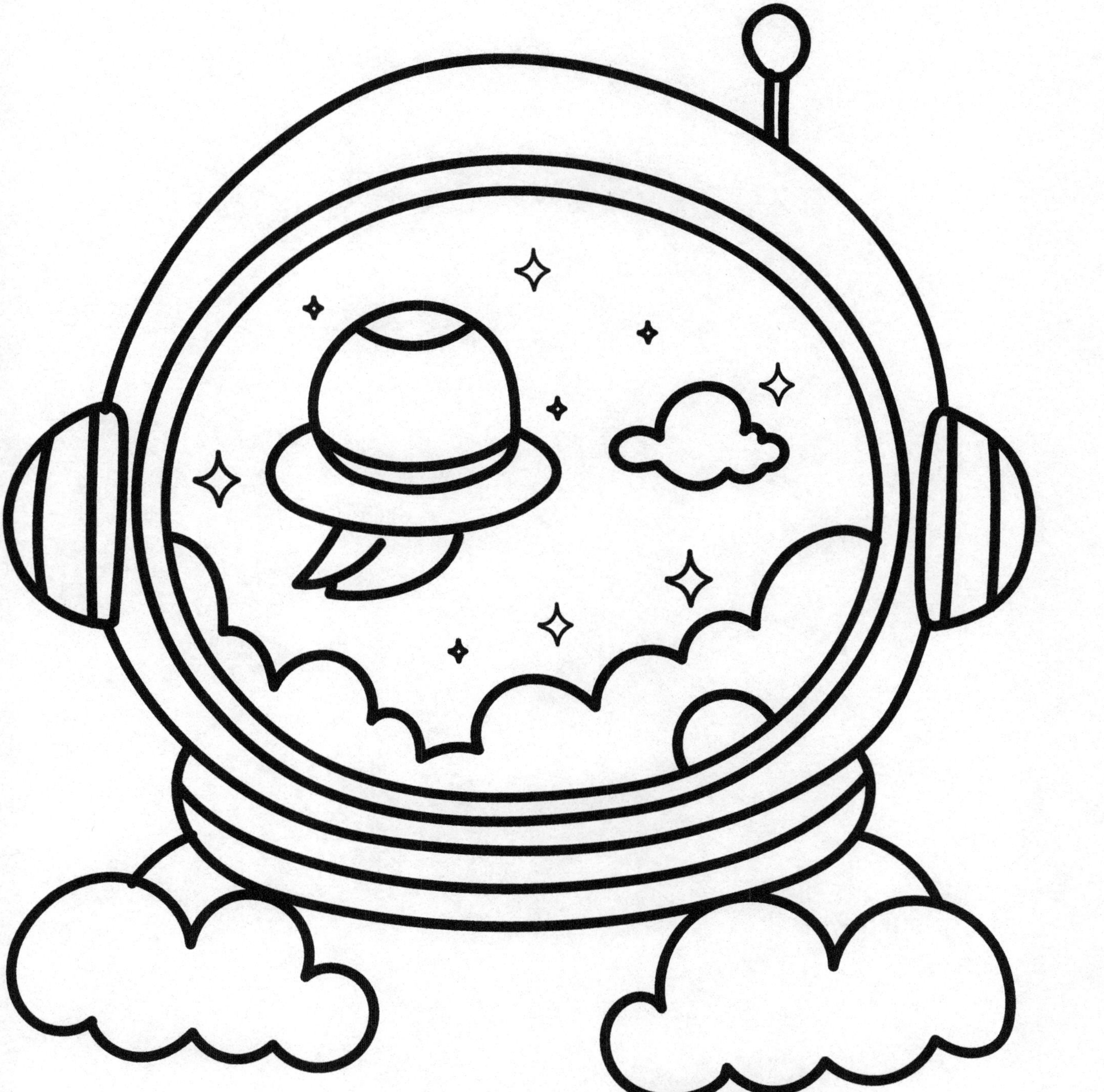

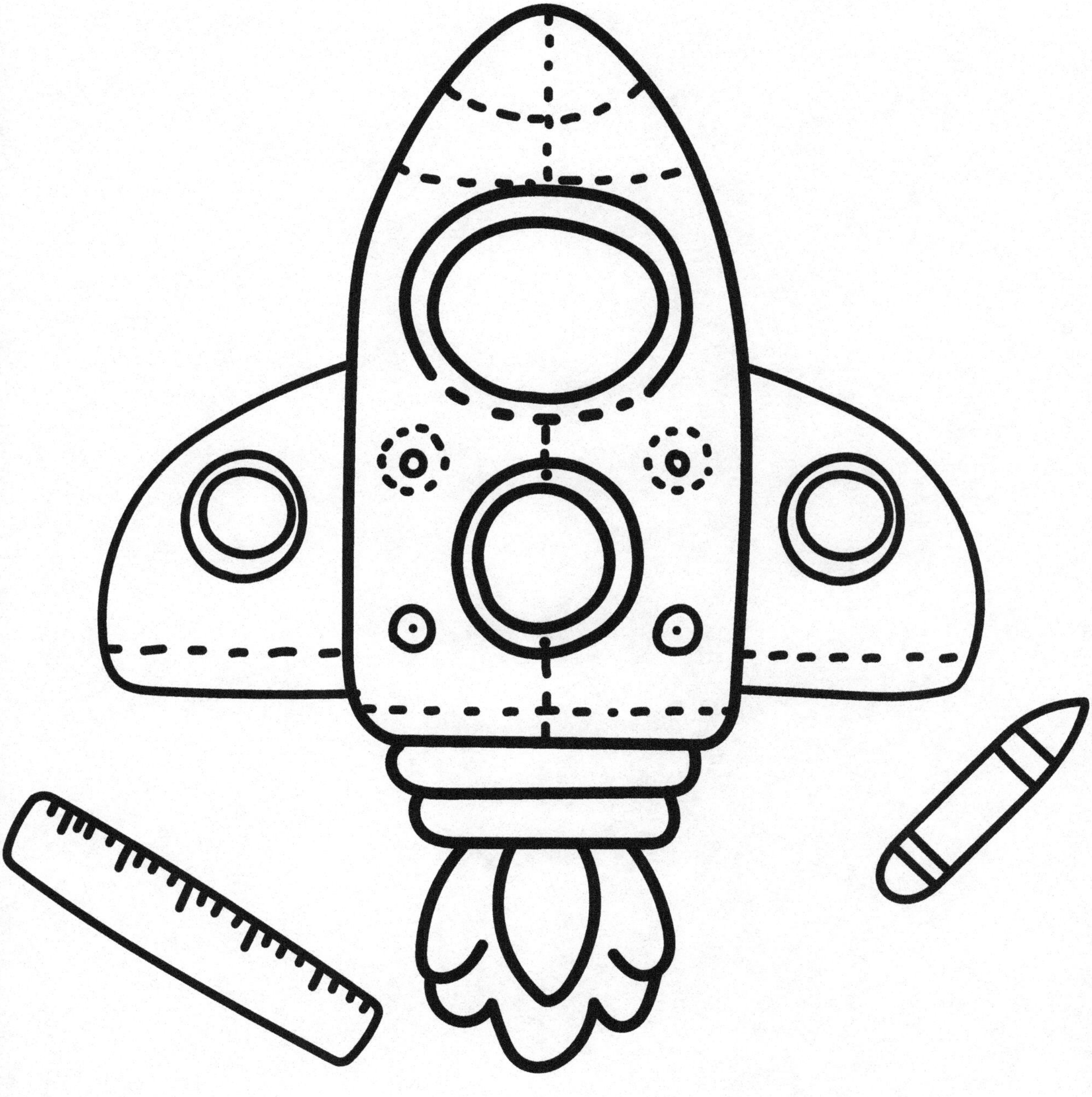

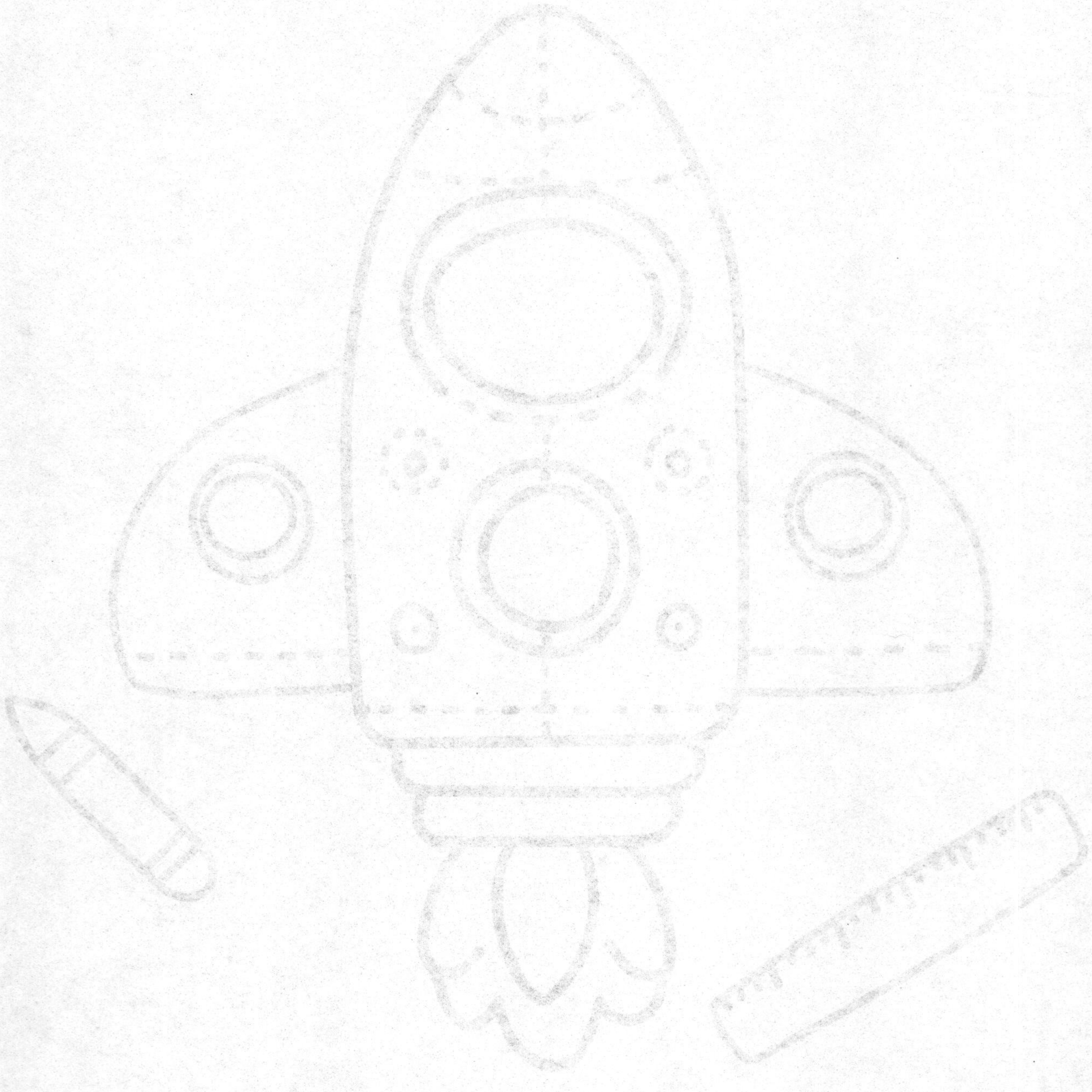

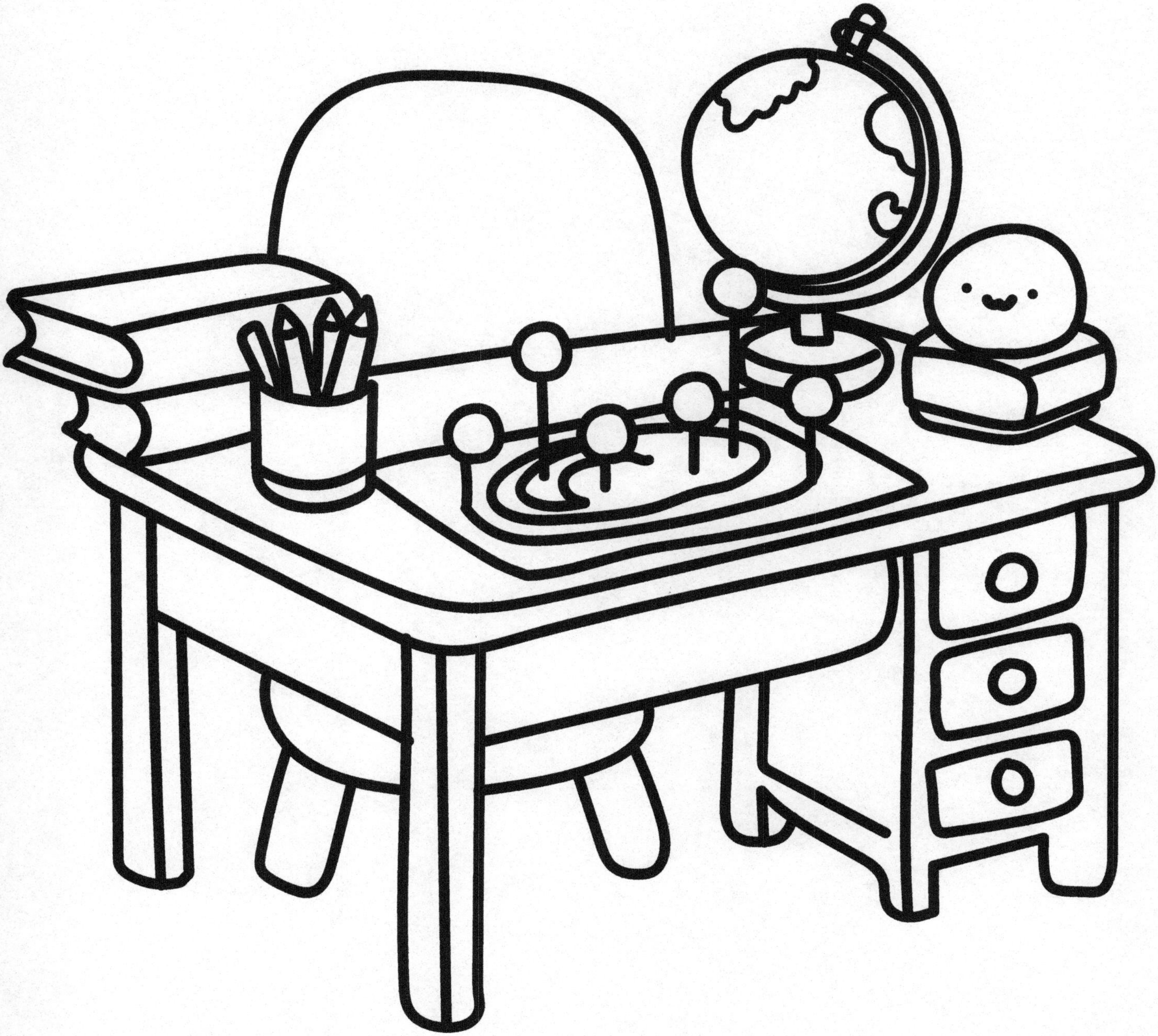

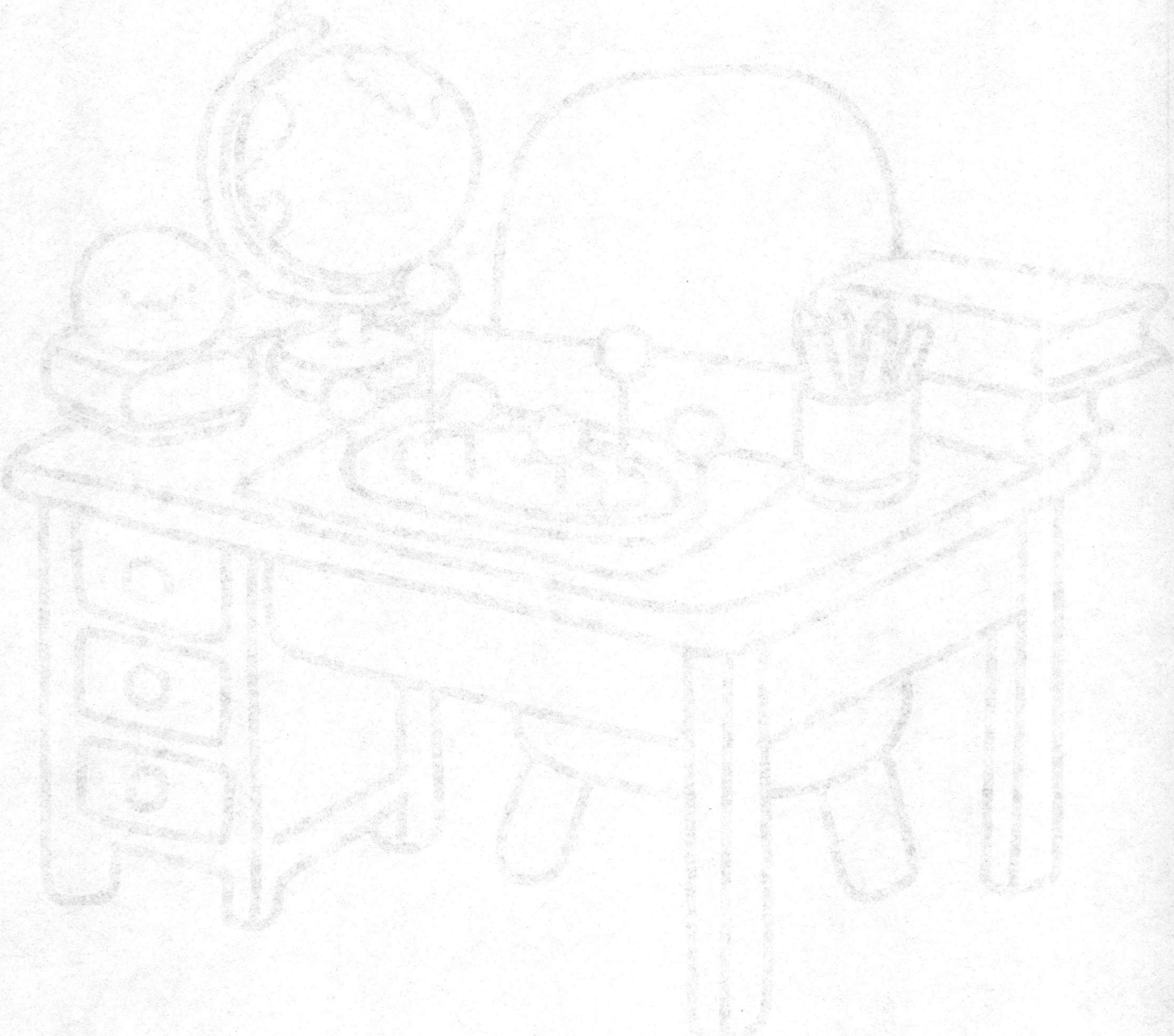

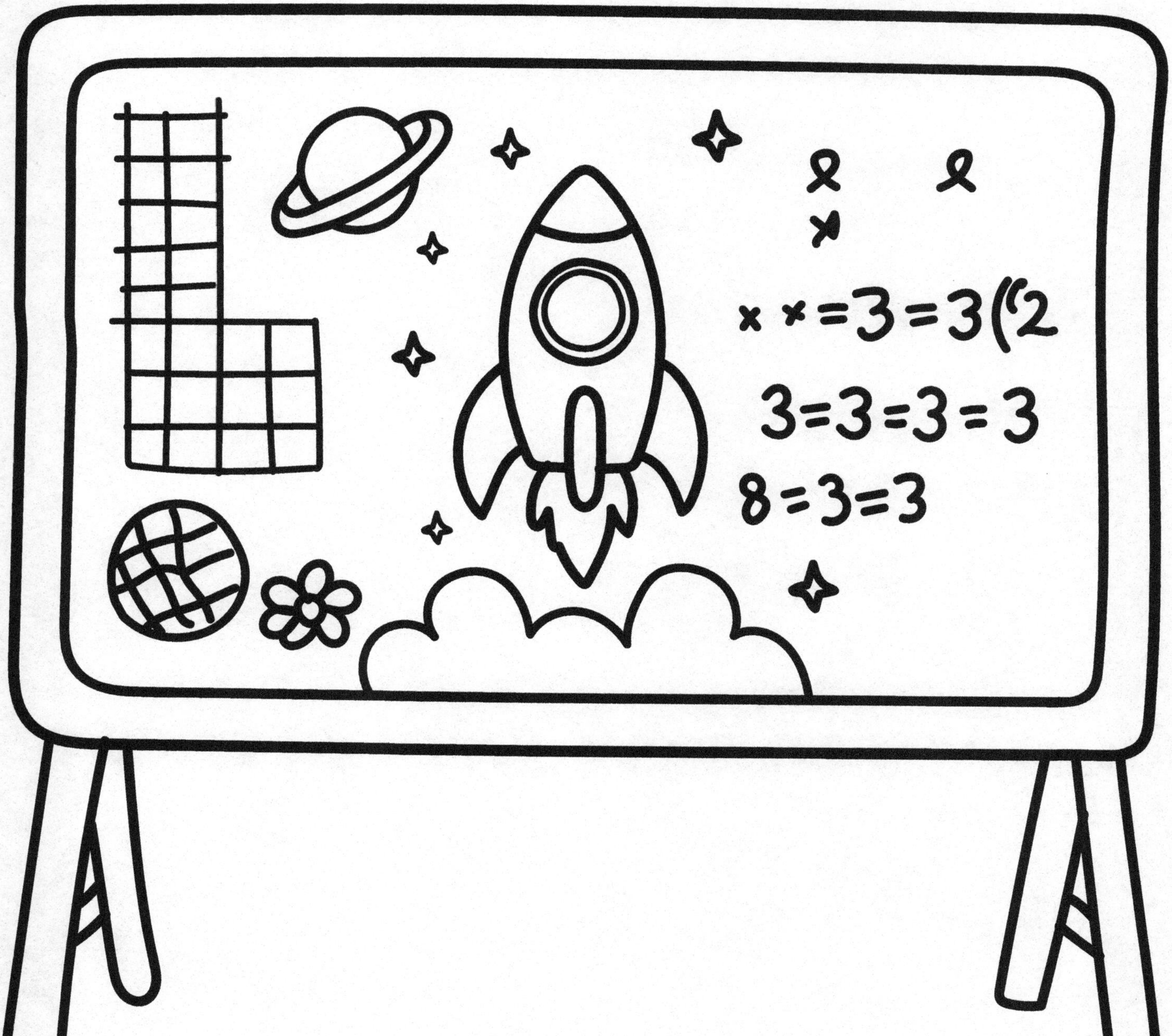
3=3=3=3
8=3=3

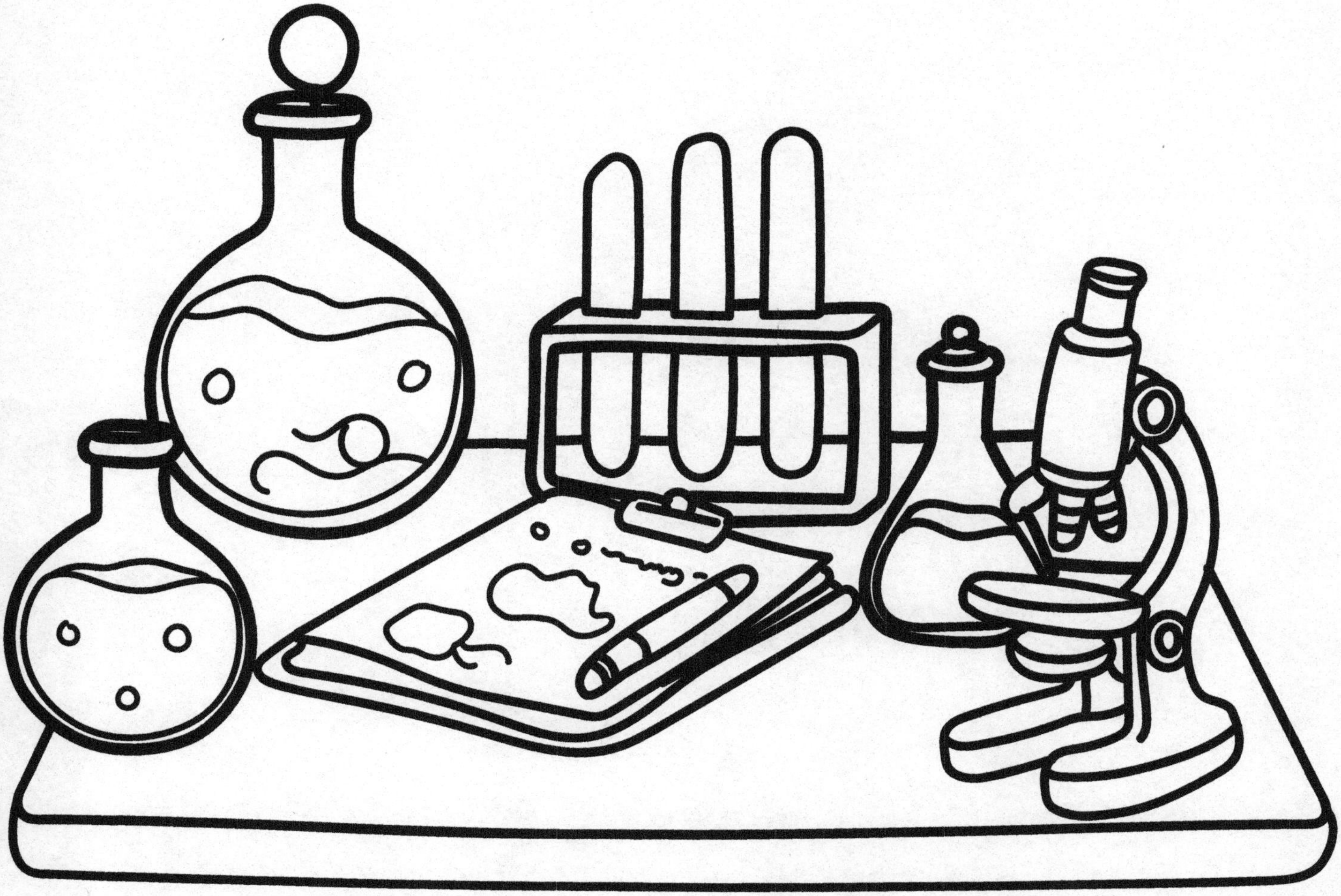

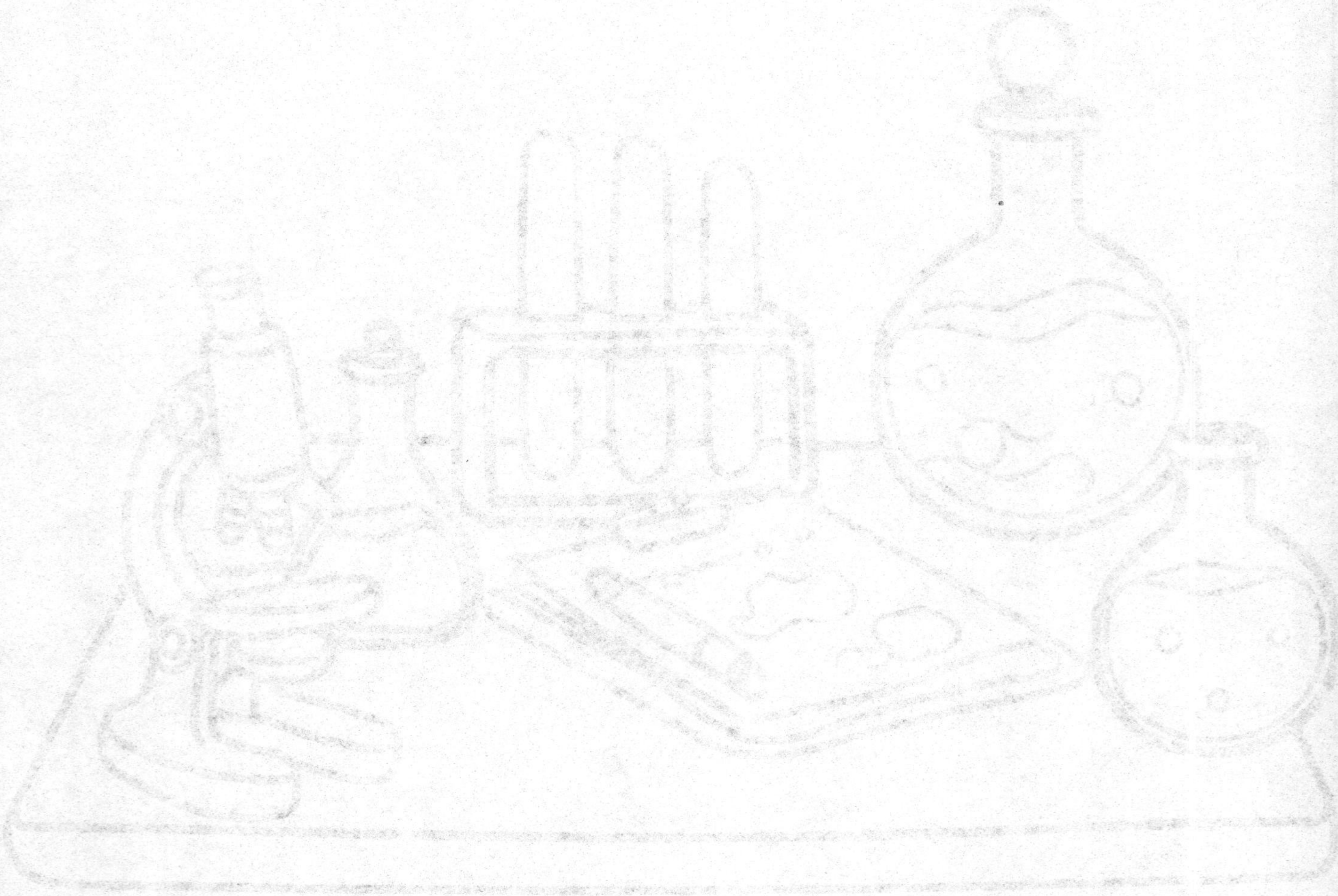

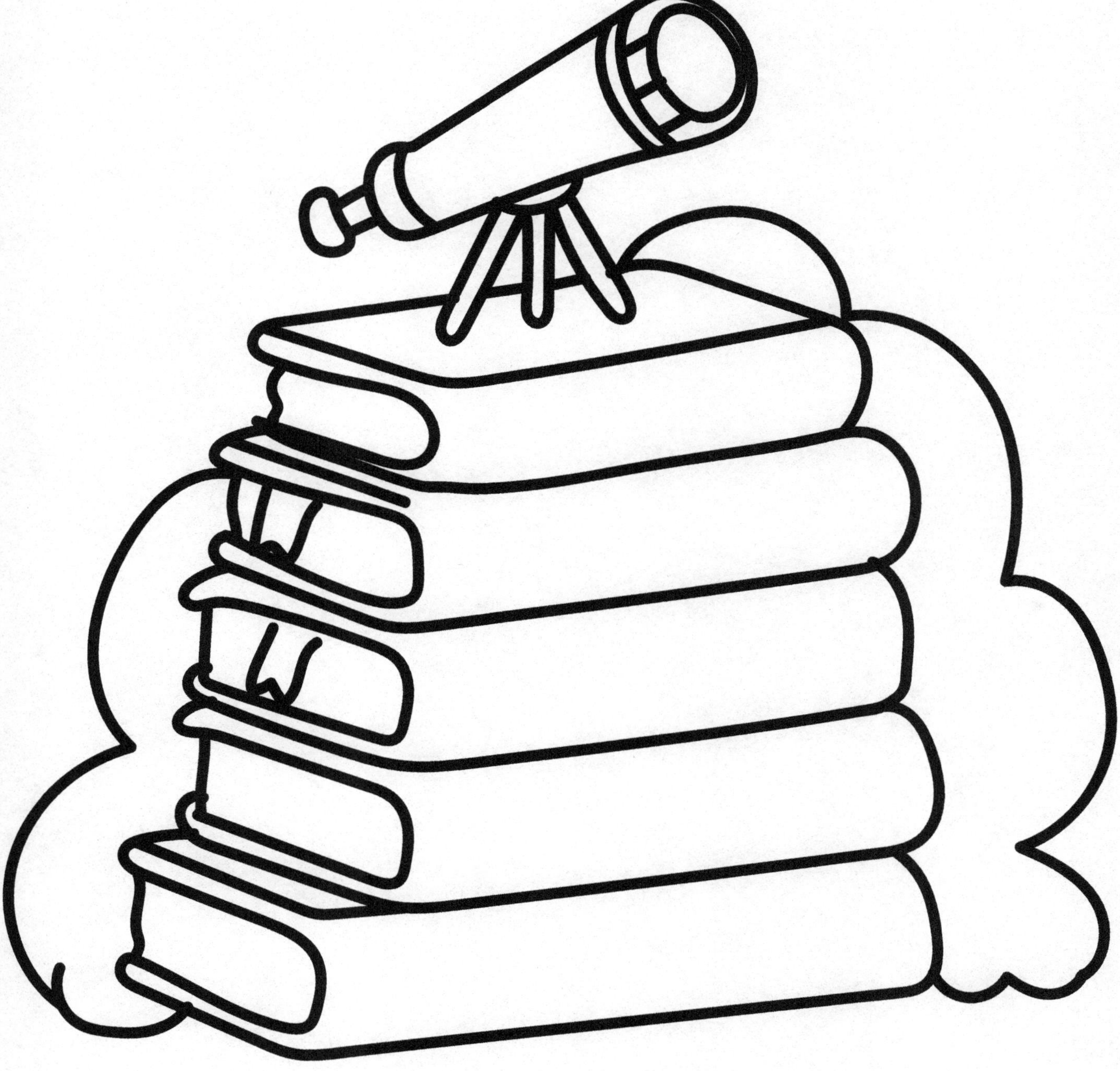

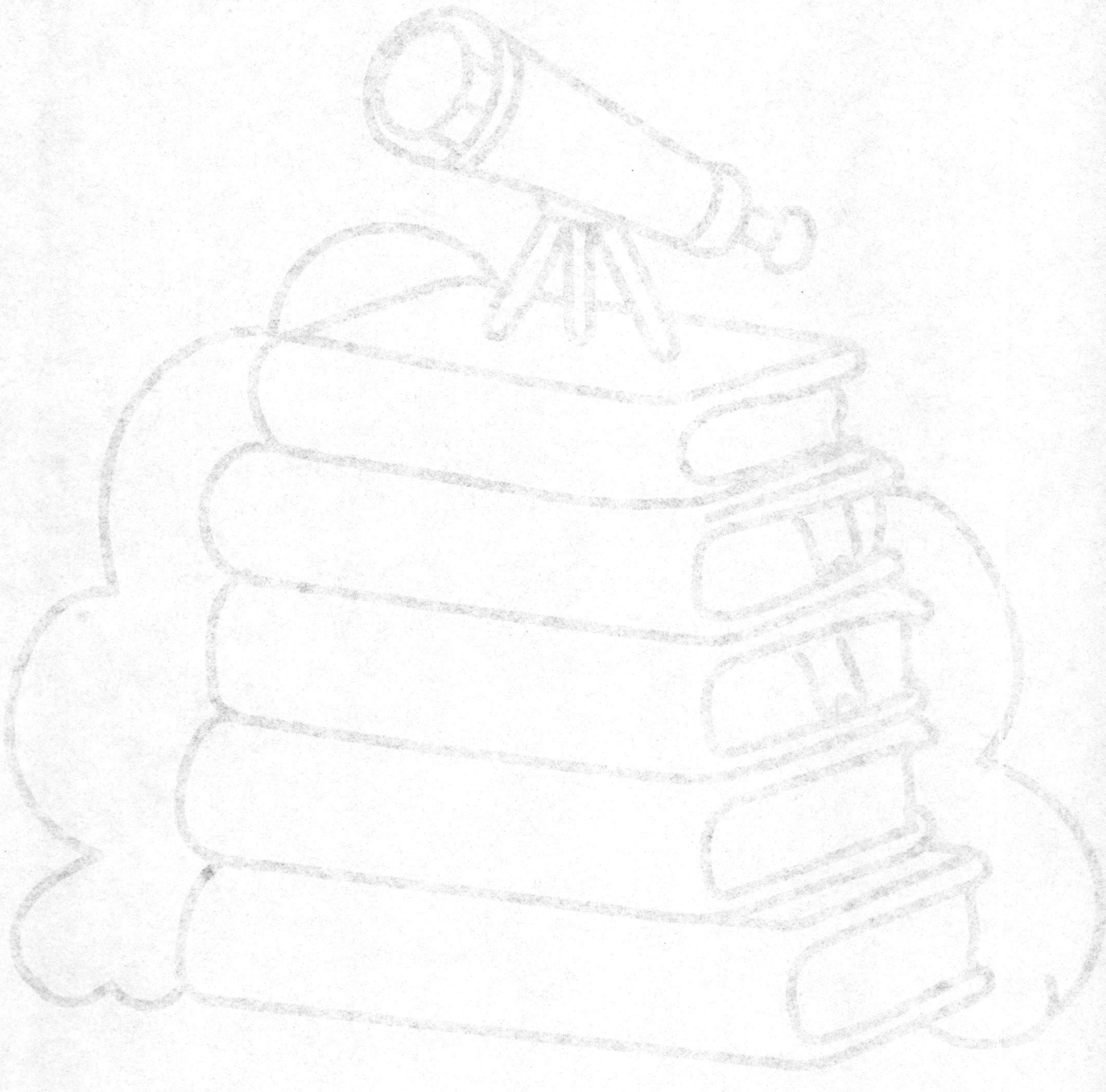

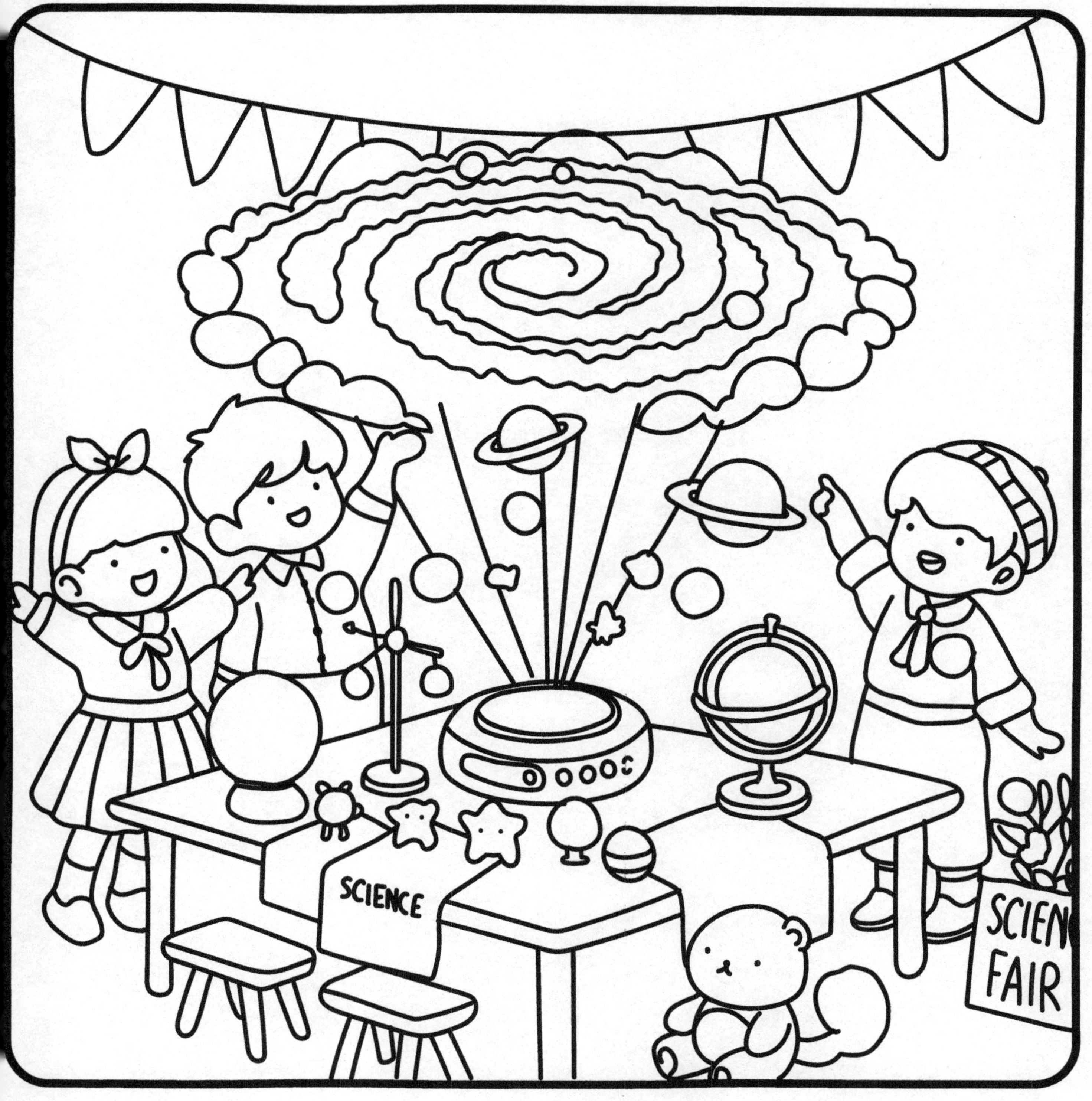
SCIENCE
SCIEN
FAIR

LOOK FOR OTHER NIA & EMMIE BOOKS

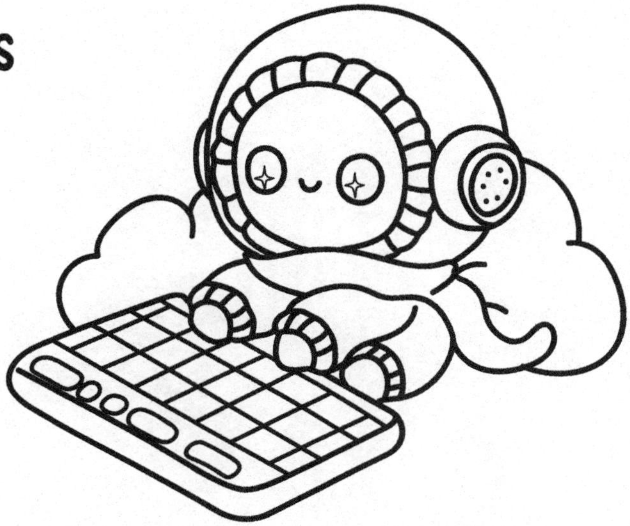

COLOR WITH NIA & EMERY:

"SWEETS & TREATS"
"FAB FASHIONS"
"FAMILY, FOOD & FUN"
"ANIMALS & NATURE"
"COMFY, COZY & CUDDLY THINGS"
"ACTS OF KINDNESS"
"PLANES, TRAINS, TRUCKS & CARS"

THE PRETTIEST GIRL IN THE WHOLE WIDE WORLD (AGES 6-9 YEARS)

In this heartwarming adventure, Nia and Emery learn an important lesson about what makes a girl truly beautiful—from the inside out. Follow their journey as they grow in love and respect for each other with a little guidance from Mama.

Available in Paperback, E-book, and Audiobook Where Books Are Sold!

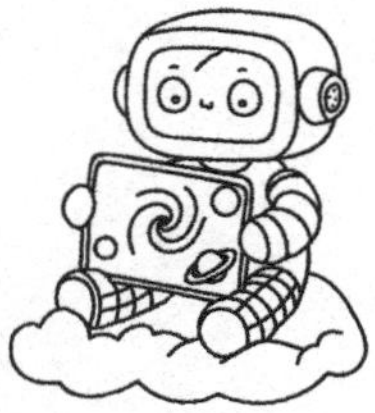

NIA & EMERY'S ULTIMATE TRIVIA CHALLENGE: JOKES, FUN FACTS, AND BRAIN TEASERS (AGES 6-9 YEARS)

Packed with brain-busting questions, giggle-worthy jokes, and clever brain teasers, this trivia book is a fun-filled way to fuel your child's curiosity and love for learning. Join Nia & Emery as they lead your child through a world of fascinating facts and silly riddles.

Available Where Books Are Sold!

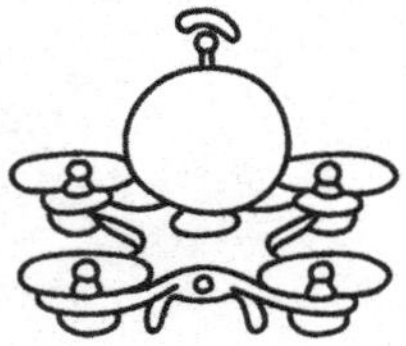

10 THINGS WE DIDN'T KNOW ABOUT JESUS! (AGES 5-9 YEARS)

Discover 10 interesting and encouraging facts about Jesus of Nazareth, presented in a child-friendly way. Your little one will learn about His love, miracles, and even His cooking skills, experiencing the wonder and love of the world's most influential figure in history.

Available in Paperback, E-book, and Audiobook Where Books Are Sold!

Made in the USA
Monee, IL
08 November 2026

68002513R00063